Planned Grazing: A Study Guide and Field Manual

For use with the book

Holistic Management: A New Frame Work for Decision Making

by

Allan Savory with Jody Butterfield

And its companion handbook

Holistic Management Handbook: Healthy Land, Healthy Profits

by

Jody Butterfield, Sam Bingham and Alan Savory

Compiled and Edited

by

Dr. Jimmy T. (Gunny) LaBaume, PhD, ChFC
President & CEO
Land & Livestock International, Inc.
Managing the Ranch as a Business™ with Restoration Grazing™
904 West Avenue H
Alpine, TX 79830
Web Site: www.landandlivestockinternational.com
Blog: www.landandlivestock.wordpress.com
E-mail: drjimmytlabaume@landandlivestockinternational.com

Preface

From the Editor: I first attended Alan Savory and Stan Parson's "Rancher School" in Albuquerque, NM in the early 1980s. Since that time, and throughout my academic career, I have advocated, taught and practiced the principles of the "Savory Grazing Method (or SGM)" as it was called in those days.

Only a short time after I completed the school, Alan and Stan decided to dissolve their partnership and go their separate ways. Alan founded Holistic Management International and Stan the Ranching for Profit School.

SGM (now Holistic Management) is a "method" and "methods" cannot be patented. As a result, the names the method is presented under have proliferated. Although the fundamental ecological principles have not changed since those early days, we now have managed grazing, planned grazing, mob grazing, management intensive grazing, and perhaps a half-dozen others. (In fact, for some small degree of distinction, we choose to call what we do "Managing the Ranch as a Business with Restoration Grazing.") Never-the-less, to this day I still find myself referring to it as "The Savory Grazing Method" in honor of the brilliant man from whose mind these ideas sprung.

This guide is a detailed book review of sorts, but perhaps it would be more accurately described as an abstract that condenses 864 pages of detail into 190 pages of concentrated information. First, *Holistic Management* by Alan Savory, et al and *Holistic Management Handbook* by Jody Butterfield, et al were condensed and summarized. Then, the summaries were complied into one and the result organized into a logical sequence.

In no way is this booklet intended to be a substitute for the original works. They are both highly recommended and readily available from Land & Livestock International, Inc., Holistic Management International, Amazon.com and many other sources.

Suggested uses might include:
- lesson plan guidelines for educators
- a study guide for students and
- a field guide and quick reference manual for practitioners

The material is organized into nine "Lessons:"

Lesson 1: Introduction
Lesson 2: The Strategic Goal
Lesson 3: The Four Fundamental Ecosystem Processes
Lesson 4: Tools for Ecosystem Management
Lesson 5: Management Guidelines
Lesson 6: Infrastructure Planning
Lesson 7: Grazing Planning
Lesson 8: Financial Planning

Lesson 9: Monitoring and Control

In sum: the primary objective was to capture the essence of the two books in as few words as possible and in a language understandable by the reasonably intelligent person without formal education in natural resource or business management.

My hope is to make these logical, common sense ideas available to a wider audience and, thereby, make a small contribution to the conservation of the world's natural resources. If I have succeeded, that would be my only small claim to originality because the ideas are from the minds and the pens of Allan Savory, Jody Butterfield and Sam Bingham. The errors are mine.

A few comments on the terminology: Those even vaguely familiar with the "holistic" approach to planned grazing will immediately recognize that I have avoided the use of certain "buzz" words that, frankly, offend me and for good reason.

Take, for example, the words holistic and sustainability. By dictionary definition, both words have meanings to which hardly anyone could object. Who in their right mind could be opposed to sustainable agriculture? Especially when dealing with human and natural systems, very few would deny the concept that the whole is greater than the sum of its parts or that the organization's land, its people and their money should be viewed as one.

Never-the-less, these words have been co-opted by malevolent and misguided elements in society and incorporated into a code that furthers their agenda. This terminology has become the language of radical environmentalism as advocated by the United Nations and its Agenda 21.

This massive wealth re-distribution scheme is one of the most destructive forces of life, liberty and property ever faced by mankind. Their claims that overpopulation, declining energy resources, deforestation, species loss, water shortages, certain aspects of global warming, and an assortment of other global environmental issues are unsupported by analysis of the relevant data.

Where" Holistic" Management Crosses Paths with Agenda 21: The anti-private property and anti-free market "sustainable development" movement utilizes triple-bottom-line accounting (TBL). TBL was created by the United Nations to advance the four main initiatives birthed at the 1992 U.N. Earth Summit in Rio de Janerio: Climate Change, Agenda 21.

TBL (also known as people, planet, profit or 'the three pillars') captures an expanded spectrum of values and criteria for measuring organizational (and societal) success: economic, ecological and social. With the ratification of the United Nations and ICLEEI (International Council for Local Environmental Initiatives) TBL became the standard for urban and community accounting in early 2007. It has now become the dominant approach to public-sector full-cost accounting.

In the private sector, a commitment to corporate social responsibility (CSR) implies a commitment to some form of TBL reporting.

With Holistic Management, decisions are made and tested for soundness and to make sure they will take you toward your holistic goal. There are seven of these tests, including "sustainability, society and culture" – i.e. echoes of Agenda 21.

Of course, people who make their living in the free market know that economics already accounts for "society" and "environment." Every day in every purchase decision made by each of the approximately 330 million people in the United States, a value is given to society and environment through price. Imposing a value for society and environment ensures they are double counted. No matter how well-intended industry's acknowledgement of the triple bottom line, there is no escaping the fact that it sets producers up for a tax at some point in the future. Whatever extraneous values are agreed to will eventually become a financial penalty on production.

We do *not* advocate the use of triple bottom line accounting.

Neither do we agree with other aspects of the financial planning methodology for example, the contention that all planning forms should be prepared by hand. Instead, we advocate taking advantages of modern technology where it is appropriate.

Although these deviations are noted in the text, we present the method in its entirety for the sake of purity and to provide the reader with the option of doing his/her planning precisely as described.

Other (but related) differences in philosophy: Practitioners of "holistic" management tend to embrace non-profit organizations and government interventionism. We do not. We strongly advocate private property and free markets as the optimum allocator of all resources—natural, financial and managerial.

Experience tells us that no government bureaucrat can (or will) care for a resource better than the man who owns a capital interest in that resource. There is nothing that any government can do that a private property-natural law based, free market, for profit, society cannot do better.

There is no better, more efficient or effective allocator of scarce resources than the free market working through its pricing mechanism. The environmental issues we are facing today would disappear as if by magic in the absence of government interventionism.

A Final Note: Unless otherwise noted, all images figures and tables come directly from the text and handbook. The planning charts and other resources are available from Holistic Management International at http://holisticmanagement.org/

So, as Alan Savory would say: Let's get on with it.

Jimmy T. LaBaume

Table of Contents

Lesson 1: Introduction	
The Decision Making Process……………………………………………………………	11
Key Insights and Resistance to Change…………………………………………………...	11
Key Insight 1: The Whole is Greater than the Sum of its Parts………………………	11
Key Insight 2: The Brittleness Scale………………………………………………….	13
Key Insight 3: The Role of Predator and Prey………………………………………..	14
Key Insight 4: The Role of Time…………………………………….............................	15

Lesson 2: Setting a Strategic Goal and Testing Decisions against it.	
What is it, exactly, that you are managing?...	17
Forming a Strategic Goal…………………………………………………………….	18
Preface: The Statement of Purpose…………………………………………………...	18
Step 1. The Quality of Life Statement………………………………………………..	19
Step 2: Identify Forms of Production………………………………………………...	20
Step 3: Describe the Future Resource Base………………………………………...	20
Refining Your Strategic Goal………………………………………………………...	21
Testing Decisions against the Strategic Goal………………………………………...	24
Power is in the Strategic Goal……………………………………………………….	25

Lesson 3: Four Fundamental Ecosystem Processes: The Focus of Restoration Grazing™	
Process 1: The Water Cycle………………………………………………………….	27
Process 2: Community Dynamics…………………………………………………...	31
Process 3: The Mineral Cycle………………………………………………………..	35
Process 4: Energy Flow: The Engine that Pulls the Train…………………………..	36

Lesson 4: Managing the Ecosystem	
Introduction…………………………………………………………………………	41
Tool 1: Money and Labor…………………………………………………………...	41
Tool 2: Human Creativity…………………………………………………………...	41
Tool 3. Fire………………………………………………………………………….	42
Tool 4: Rest…………………………………………………………………………	45
Tool 5: Grazing……………………………………………………………………..	48
Tool 6: Animal Impact ……………………………………………………………..	52
Tool 7: Living Organisms…………………………………………………………..	56
Tool 8: Technology…………………………………………………………………	57

Lesson 5: Management Guidelines	
Guideline 1: Controlling Time……………………………………………………...	59
Guideline 2: Stock Density and Herd Effect………………………………………..	62
Guideline 3: Burning………………………………………………………………..	65
Guideline 4: Population Management……………………………………………….	67

Lesson 6: Infrastructure Planning	
Introduction to the Planning Process………………………………………………..	71
Step 1: Gathering Important Information…………………………………………...	71
Step 2: Preparing Maps and Overlays……………………………………………….	73

Step 3: Deciding on Herd and Cell Sizes…………………………………………..	73
Step 4: Prepare Maps for Planning……………………………………………..	75
Step 5: Holding the Planning Session…………………………………………….	75
Step 6: Designing the Ideal Plan……………………………………………….	76
Step 7: Implementing your Plan……………………………………………….	77
Notes, Ideas and Practical Solutions………………………………………….	77

Lesson 7: Grazing Planning	
General Principles of Planned Grazing………………………………………..	93
Recap, Review and More Basics………………………………………………	99
Practical Application Creating the Grazing Plan………………………………….	111
Checklist: The Open Ended Plan……………………………………………...	112
Step 1. Make Initial Decisions: Look at the Big Picture……………………….	116
Step 2. Set Up the Grazing Chart…………………………………………….	117
Step 3. Record Management Concerns Affecting the Whole Cell……………….	117
Step 4. Record Herd Information…………………………………………….	117
Step 5. Record Livestock Exclusion Periods…………………………………..	117
Step 6. Open-Ended Plan Only: Check for Unfavorable Grazing Patterns……….	118
Step 7. Record Paddocks Still Available………………………………………	118
Step 8. Note Paddocks Requiring Special Attention…………………………..	118
Step 9. Rate Paddock Productivity………………………………………….	118
Step 10. Open-Ended Plan Only: Determine the Length of Recovery Periods………	119
Step 11. Open-Ended Plan Only: Calculate Grazing Periods…………………….	119
Step 12. Closed Plan Only: Assess Forage Volume, Carrying Capacity and Drought Reserve………………………………………………………………..	121
Step 13: Closed Plan Only: Plan the Number of Selections and the Grazing Periods……	121
Step 14. Plot the Grazings………………………………………………………..	121
Step 15. Make a Final Check of Your Plan……………………………………	122
Step 16. Implement (and Monitor) the Plan…………………………………….	122
Step 17. Record Results…………………………………………………….	123
Checklist: The Closed Plan………………………………………………….	124
A special case: Grazing two or more herds in the same cell…………………….	134
Planned Grazing Summary………………………………………………….	138

Lesson 8: Financial Planning and Monitoring	
Introduction and Overview of the Planning Process…………………………….	139
The Planning Process Itself………………………………………………….	141
Financial Planning Basics……………………………………………………	146
The Testing Guidelines……………………………………………………..	146
Financial Planning Forms…………………………………………………..	149
Step by Step Guidelines for Creating Your Plan………………………………..	157

Lesson 9: Biological Monitoring	
Introduction to the Basics of Biological Monitoring…………………………….	167
General Observations………………………………………………………	168
The Four Ecosystem Processes: Key Indicators……………………………….	169
What Grazing Patterns Can Tell You…………………………………………	172
Two Monitoring Procedures………………………………………………...	173
Gathering the Data…………………………………………………………	173

Basic Monitoring...	174
Comprehensive Monitoring..	178
Interpreting Your Results—Five Scenarios...	186
What each tool produces in terms of the 4 processes......................................	187

Lesson 1: Introduction

The Decision Making Process

Our approach to management begins with a definition of the entity to be managed in terms of the people who are to manage it and the resources they have. These people then formulate the *strategic goal* which describes the quality of life they wish to achieve and what they must produce in order to achieve it. They must also describe the future resource base that will be necessary to sustain what they must produce to provide the quality of life they want. Finally, they ensure their decisions are socially, environmentally and economically sound.

Decisions along the way are evaluated according to the same criteria they have always used. Any action taken should not only achieve its objective but also make progress toward the strategic goal. A feedback loop is set up to insure that, if the decision is not taking them where they want to go, they can correct it immediately.

The process is very similar to that used in major corporations today. The difference is that the objective is to restore deteriorating environments while simultaneously increasing productivity and profitability. Businesses of all sizes see black ink turn red if they don't consider the environmental and social consequences of their decisions. Acting responsibly is the only way to attract and keep customers. In other words, acting responsibly is very much a part of the profit motive—if you don't, customers will go away and so will the profits.

Key Insights and Resistance to Change

Even trained scientists can never be objective about new information. Over the years, Alan Savory developed four key insights that contradicted long-held beliefs about what causes environmental deterioration. Consequently, he ran into resistance to acceptance due to a lack of understanding. Savory's four key insights are: 1) The Whole is Greater than the Sum of its Parts; 2) A New Way of Viewing Environments; 3) The Role of Predator-Prey Relationships; and 4) It is all about Timing. Each is discussed below.

Key Insight 1: The Whole is Greater than the Sum of its Parts

A holistic perspective is essential because only the whole is reality. This is contrary to the long held notion that the world is like a machine with individual, separate parts that can be isolated for study or management. It is not. The world is made up of patterns that function as *wholes*. These patterns cannot be deciphered or described by studying any single part in isolation. This is why defining the whole that you are dealing with is the first step in management.

The modern scientific approach originated in the 13th century with Roger Bacon's idea which was developed into the formal scientific method. Using the scientific method one seeks to test a hypothesis by controlling all the variables of a phenomenon and manipulating them one at a time. However, when attempting to study our ecosystem and the many creatures that inhabit it, we cannot meaningfully isolate anything let alone control the variables. Isolate a single part and neither what you took nor what you left behind remains what it was when all was one.

This view was given a name by Jan Christian Smuts (1870-1950) in the 1920s--*holism* (from the Greek *holos*). Smuts attested that the world is not made up of substance, but of flexible, changing patterns—i.e. it is arrangements and not stuff that make up the world. Wholes have no stuff. They are arrangements. Furthermore, individual parts do not exist in nature, only wholes, and these form and shape each other.

Further, due to this complexity (a single teaspoon of water can contain a billion organisms), the very limitations that would make a research project acceptable scientifically also make the results meaningless.

Furthermore, approaching management from the perspective of narrow disciplines has done nothing but aggravate the problems. No animal nutritionist, soil scientist, economist, or any other specialist alone has meaningful answers. There are numerous reasons why these "interdisciplinary" teams have not worked.

First, specialists usually communicate poorly, not only because they have different perspectives, but also because they speak different languages. As a result, opinions acquire weight and conclusions are negotiated according to criteria that may have no relationship whatever to overall need.

The problem is exacerbated when the interdisciplinary team is state sponsored and therefore socialistic by nature. Back in the 1930s Ludwig von Mises identified the failure of socialism as being a result of "no means to calculate." There are no market prices which can be used to allocate scarce resources. There are no profit centers to hold anybody responsible for. Thus, that leaves egos and personal agendas in charge of decision making. Interdisciplinary teams assigned to "study" a problem are the result of government attempting to circumvent this "calculation" problem [or cover up the fact that there is such a problem] and thereby justify its own existence.

Not only are there no parts in nature, there are no boundaries either—only wholes within wholes in a variety of patterns.

Scientific custom isolated the individual parts for study believing that if we could learn enough about each of them we would understand the whole. However, in nature this leads to nowhere. But of course, once you see the whole in the pattern, detailed knowledge of the individual parts does become useful. However, one can ask the right questions about the details only by having first seen the whole.

The fact that wholes have qualities not present in their parts causes the interdisciplinary approach to flounder.

All management decisions have to be made from the perspective of the whole. Therefore, first the whole must be defined keeping in mind that it is always influenced by both greater and lesser wholes. Then, in order to know what we want to do with it, we needed an all encompassing strategic goal. Then, we also need a means of weighing up the ramifications of our actions.

Given the complexity of the world, computers are more capable of weighing up the consequences of a particular decision. They can be powerful tools for solving specific mechanical problems. However, they cannot think holistically or evaluate emotions and human values which are vital components of the whole. On the other hand, the human mind can see patterns and make decisions out of a deep, even unconscious, sense of the whole. This is the function of a "free" market. No individual can possibly know everything there is to know at any given time but the "free market" can (and does) because it is the sum total of human wisdom at that given time.

Advice that appears perfectly sound from an economic or engineering or any other view is likely to be unsound holistically. This has spelled disaster for many a foreign aid project and national policy, but also for families, communities, and businesses large and small. Most, if not all, such disasters are a direct result of government "intervention" – people do not make policy decisions. Governments do. Then government forces the people to accept the decision at the point of a gun. Many a private rancher has gone bankrupt after committing scarce funds to government-sponsored irrigation schemes. There are always risks to tying up capital in government incentive programs.

Key Insight 2: The Brittleness Scale

Environments can be classified on a continuum. This classification is based on how moisture is distributed throughout the year and how quickly dead vegetation breaks down. On opposite ends of this continuum are two broad categories of environments (brittle and non-brittle) that will respond differently to the same treatments. For example: Rest will restore non-brittle land but will damage it in very brittle environments.

This second insight flies in the face of the belief that all environments respond in the same manner to the same influences. They don't.

It has been long recognized that some environments readily deteriorated under human management. The world's arid and semi-arid areas are mostly some type of grassland with livestock production being the primary economic activity. When livestock management produced bare ground, moisture either evaporated or ran off. Common sense dictated that the best remedy is to rest the land from human disturbance. That would seem to be a logical assumption.

However, Savory's experience in the Rhodesian game department led to his suspecting a fundamental flaw in this belief. He could not explain what he saw because it did not fit the neat scientific theories. He *knew* that fire maintained grassland. The only other influence he *knew* that could cause such damage was overgrazing. Yet, when the game populations were decimated and the incidence of fire increased, the land actually deteriorated. Neither game nor domestic animals were present—there was no overgrazing.

He once published a paper in which he concluded that, once land was so badly damaged, it reached a point of no return and would never recover. It was years later before he realized how wrong he had been.

What had eluded him was that there are two broad types of environments that react differently to management. Practices that benefit one damage the other. It was from this insight that he coined the terms "brittle" and "non-brittle" with true jungle being a 1 on the brittleness scale and true deserts being a 10. Everything else is somewhere in between.

Brittleness is not the same as fragility. You can actually have a fragile community within a non-brittle environment (such as a delicate fern glade within a forest).

Brittleness is derived, not so much from total rainfall as from the distribution of moisture over the year. Brittle environments are typically characterized by erratic precipitation and humidity during the year. For example, a 30 to 50 inch rainfall area that has a very dry period in the middle of the growing season is likely to be very brittle. On the other hand, a 20 inch rainfall zone where the moisture reliably comes during the growing season would be non-brittle. In a completely non-brittle environment, precipitation and humidity would be constant and high.

The poorer the distribution of moisture, particularly in the growing season, the more brittle the area tends to be, even though total rainfall may be high. Brittle environments commonly have a long period of non-growth that can be very arid.

Where moisture is erratic (brittle environment), vegetation, insect, and microorganism populations build up during the rainy months. However, when the rain stops, the vegetation dies and the insect and microorganism populations decline drastically. The plant's stems and leaves are dead and of no use to the plant. In fact, this remaining material is a liability in that it blocks sunlight from the ground level growth points of the plants.

For millions of years in the past, lightning had some influence but this occurred over relatively few areas in any one year. The difference is that there were lots of grazing animals that continued to consume the plant material long after growth had stopped. The microorganisms in their digestive tracts reduced the material to dung. In the following rainy season, insects and microorganisms became numerous and consumed the dung and dead vegetation that had been trampled onto the soil.

Today, there are no more vast herds. As a result, only a small portion of the vegetation produced is able to decay. Thus, the old belief that *all* land should be rested is wrong. Although non-brittle environments do, in fact, respond to rest, in brittle environments rest will lead to further deterioration and instability.

Bare ground is another indicator of brittleness. In non-brittle environments it is difficult, if not impossible, to create large areas of bare ground whereas in brittle environments bare ground is easily created.

In sum, different environments respond differently to the same influences. Rest restores non-brittle environments but damages brittle ones. The answer to how grazing animals might provide the necessary disturbance lies in the remaining two insights.

Key Insight 3: The Role of Predator and Prey

The types of animals associated with the two types of environments are also different. A relatively large number of large bodied herd animals, naturally concentrated and moving in the presence of predators, are vital to maintaining brittle environments. Much of the deterioration was initiated by the severing of the essential relationship between herding animals and their pack-hunting predators.

Animals behave differently in different situations. For example, large buffalo herds spread out when feeding, but not too far. They walk gently and slowly placing their hooves beside coarse plants and not on top of them. They also place full weight on the hooves compacting the soil. They make little impact on the plants and the soil other than the removal of forage and soil compaction.

However, once the herd begins to move or when predators are around, they bunch up as a herd for safety. They step recklessly and trample down very coarse plants containing old material providing cover for the soil surface. Their hooves leave the soil chipped and broken. In other words, the animals do what any gardener does. They first loosen the sealed soil surface, then bury the seed slightly, compact the soil around the seed and then cover the surface with mulch.

Even where they are herded (as opposed to being fenced in), they do not behave as they would under the threat of predation.

Large predators behave differently in different environments. In brittle environments wolves, lions, wild dogs, etc. hunt in packs and run down their prey. On the other hand, in non-brittle environments the predators (tigers, jaguars, etc) hunt singly and ambush their prey. Their counterparts in the more brittle environments (leopards and mountain lions) do not associate with large herds.

It was the pack-hunting predators who were mainly responsible for producing the change in behavior of their herding prey. Bunching up was the herd's main form of protection.

Especially in brittle environments, large, concentrated and moving herds of animals in the presence of pack-hunting predators are vital to the health of the land.

But historically, as bare ground increased and the environment deteriorated, we attributed it to overgrazing which we, in turn, blamed on too many animals. We decreased animal numbers and thus increased the bare ground and the deterioration.

The fourth insight is that overgrazing is not in fact a function of animal numbers.

Key Insight 4: The Role of Time

Damage from overgrazing and trampling has little relationship to animal numbers. Instead, the important factor is the time that plants and soils are exposed to the animals.

Traditionally, preventing overgrazing has always begun with limiting livestock numbers. The conventional recommendation has been to regulate numbers so that animals do not graze more than half of certain "key" species.

This take half-leave half theory was doomed to fail from the beginning. Animals simply do not graze individual plants that way. There will be some plants that won't be grazed at all while others are grazed right down to their growing points at the base. So, the land continues to deteriorate under this approach to range management.

Meantime, conventional range managers judged their success by the presence of a few "desirable" species and the general appearance of the land. But, even their research plots are developing bare soil between the plants and showing signs of erosion. Some plants continue to be overgrazed in patches while others grow old and fibrous from being smothered by accumulations of oxidizing material.

So, conventional range managers attempt to solve the uneven grazing problem with periodic burning while burned areas continue to visibly erode. Very few species are present other than the so-called "desirables." There is an almost complete absence of new plant seedlings. Production per animal is relatively high but production per acre is low and declining.

Back in the late 1950s French researcher Andre Voisin established that there was very little relationship between overgrazing and animal numbers. Instead, the primary factor was the time plants were exposed to the animals. Meantime, the "excessive animal numbers cause overgrazing" type of thinking continues unabated.

Voisin discovered that the critical time of exposure was determined by the growth rate of the plant. If the plants are growing fast, the animals need to be moved on more quickly and vice versa for periods of slow (or no) growth. And, in addition to grazing, trampling could be either good or bad depending on time.

Then Alan Savory reasoned out how time was involved in the grazing and trampling of wild herds of the past. Wild grazers bunch closely for protection from predators. They foul the ground with dung and urine. Since no animal likes to feed after itself, they kept on moving to fresh ground and would not return to the fouled area until the dung and urine had weathered and worn off.

First attempts to duplicate what the wild grazers did involved concentrating domestic animals so they were forced to eat all plants evenly. These went by names like Short-Duration Grazing or High Intensity-Low Frequency Grazing. Livestock condition suffered.

Later, Savory solved the over-grazing side by side with under-grazing phenomenon by combining the ideas of concentrating animals but not forcing them to graze plants evenly. This is accomplished by controlling the time that plants are exposed to the animals according to the rate at which the plants are growing.

He also has shown us that under-stocking (low livestock numbers) damages the land as much (sometimes even more) than over-stocking. It does this by causing perennial grasses to die from over-rest as old accumulated material blocks sunlight. In addition, too few animals scattered widely over the land do not provide the soil disturbance necessary for a healthy, fully productive range.

Lesson 2: Setting a Strategic Goal and Testing Decisions against it.

What is it, exactly, that you are managing?

How do you define a whole if wholes have no defined limits?

There may be no limits but there is a minimum which includes the people directly involved, the resources available (physical assets as well as people who can assist), and the money on hand or that can be generated. Start with an arbitrary definition of the entity you manage—a bakery or a ranch, for example.

In the process of defining the boundaries, you are identifying who will form the goal and what they will be managing with full cognizance that any whole includes lesser wholes and also lies within greater wholes.

A minimum manageable whole is made up of three parts: 1) the decision makers, 2) the resource base and 3) the money.

The Decision Makers: Decision makers include anyone that makes day-to-day decisions ranging from the profound, far-reaching to the mundane. Also include people who, while not making decisions, can veto or alter them in some way.

The Resource Base: List the major physical resources from which you will generate revenue or derive support for achieving your goal. These need not be owned, but must be available.

Then make a list of all the people who will (or can) influence (or be influenced by) management decisions but won't have the power to veto or alter them—clients, customers, suppliers, etc. These people are no less important than the decision makers. Even though they do not make management decisions, their views and concerns and how you want to relate to them should be reflected in your goal. Many, although not forming the holistic goal with you, will prove helpful in achieving it. Some, like customers or clients, will be essential.

Money: List the sources of money available to you—cash on hand, savings, relatives, shareholders, a line of credit at the bank—and the money that could be generated from the resources listed in your resource base. Think in terms of what you require to live on.

Keep Your Focus on the Big Picture: At this stage you need not concern yourself with details because, if you do, you could lose sight of the whole. Try to keep your lists and notes brief. At this point all that is needed is big-picture clarity.

Including the Right People in the Right Place: In the early days of what was then known as "the Savory Grazing Method," two eight-year-long trials were conducted by the uS Forest Service in cooperation with two ranching families. Only the families were included in the initial phase because they were the ones directly involved. The Forest Service people were included in the resource base.

Neither trial succeeded. The primary reason was that the FS people were not included in the initial phase but they had veto power over major decisions. Consequently, uSFS regulations overrode critical decisions and made it impossible to succeed.

Subsequently, agency personnel were included in the formation of the holistic goal. Then, when decisions were made that could have been vetoed, they worked to find ways around their own regulations because they were standing in *their* way too.

Don't put off defining the whole because you are afraid you won't get it right. You will make mistakes but should be able to rectify them before any serious damage is done.

Wholes within Wholes: If the group of people is very large and/or enterprises very diverse or if members are widely separated geographically, it might be impractical to manage the entity as a single whole. In some cases it would make more sense to create smaller, more manageable wholes within the greater whole. Each of these smaller wholes would meet the minimum whole requirement.

In the large industrial corporation people might number in the thousands. Each factory or division would have people directly involved in its management. Each would be fairly distinct with its own clients and suppliers. Each would have money. Thus, each factory or division could be a smaller whole within the greater whole.

Decision makers in each of the smaller wholes would form their own holistic goal to address their needs, desires, and responsibilities yet still be in line with the holistic goal formed by those managing the greater whole.

If you are managing business and family as a single whole (as many ranchers do), you may need to define two wholes—especially if some family members do not wish to be involved in the business.

The purpose in creating separate wholes is to give the people involved the opportunity to form a strategic goal that relates to their specific management needs and the resources available to them. *The more specific the strategic goal, the greater the commitment of the people involved.*

Forming a Strategic Goal

The goal should reflect the values of the people involved (spiritual and otherwise)—the things they live for, the things that make them want to do anything.

In order to gain the personal commitment needed to achieve whatever you have to achieve, it should include a quality of life statement that reflects what is most important to you.

At that point you begin to understand what you have to produce to create the outcome you envision. Then, once you know that, you can begin to envision the sort of landscape that would sustain what you need to produce.

In other words, there is an order to the formation of a strategic goal. Each aspect naturally leads to the next. The quality of life, forms of production and future resource base are combined into one comprehensive strategic goal.

The strategic goal embraces human values and links them as one indivisible entity to economies and the environment. Sharply defined objectives have their place once a holistic goal has been formed.

Weighing all decisions against this goal increases the chance of success.

Preface: The Statement of Purpose: A statement of purpose is the preface to your strategic goal.

If the entity you manage was formed for a specific purpose that you are legally or morally obligated to meet, your strategic goal will need to address that.

If the statement of purpose takes more than a sentence or two, you have either: 1) not thought carefully enough or 2) gone beyond a statement of purpose into *how* you see yourselves doing whatever it is that you are supposed to do.

The statement of purpose will be reflected in the quality of life statement of your strategic goal (where you refer to the outcomes that correspond to your purpose) and also in the forms of production (where you specify what you must produce to insure those outcomes).

Step 1. The Quality of Life Statement:

In most cases the entity you are managing will include more than one decision maker. So, the quality of life statement must become an expression of the desires and aspirations of all the decision makers—a reflection of your shared values.

The quality of life portion expresses the reasons you are doing what you are doing—what motivates you, what excites you. You must develop a collective sense of what is important and why.

Getting Down to What You Value Most: What needs to be included is unique to each situation and the values of the people involved. However, there are four areas that are commonly considered.

Economic Well-Being: Of course at least some level of economic well-being is essential for basic human needs. However, a statement such as "making a lot of money" is rarely as useful as a statement of what you gain from having money—e.g. security, comfortable surroundings, enough to eat, and the wherewithal to do what you want to do. In other words, express what you hope to gain from a thing, rather than the thing itself: Why do you want to own a game ranch? What are your family's needs? In general, what kind of life you are seeking?

Relationships: People seldom function as well when they feel alone or alienated as they do when they feel like they belong. You need to develop a collective sense of what you expect to give and receive from your relationships.

Challenge and Growth: Humans need challenge or else they fail to grow and develop. To find the challenge, think in terms of what you find stimulating. What kind of atmosphere might you create to keep everyone enthusiastic but not overwhelmed?

Purpose and Contribution: People give their best effort only when it has meaning for them. Meaning cannot be created by a leader and handed down. It has to be a shared discovery.

In our case we are working to advance rangeland restoration and managing the ranch as a business. But what really inspires us is the world we envision in which people live well, grazing lands are healing, wildlife is abundant and the fishing is good.

There is a reason why any group of people work together. What is it that you can achieve collectively that you could not achieve individually? You are trying to form an idea of what you contribute to the world.

Creating a quality of life statement may take months or even a year or more. So start with a rough statement that indicates the general direction you want to go. Then you can form the remaining two parts of your holistic goal and start making decisions that lead you toward it.

Crafting Your Statement: Capture your thoughts in simple phrases. Resist breaking the phrases down into the values they represent. If you don't, all you will end up with is a long list of values with no context.

Don't just write: "prosperity, security, family values and health" when what you really want is "to have stable and healthy families where all generations feel secure and are cared for."

Generally, people want many of the same things, regardless of their position. However, specifics may vary and in some cases conflict. Usually each can accommodate the other as long as the difference has not already caused conflict and hurt.

Accommodate, rather than compromise. Anyone forced to compromise will not have much commitment to the strategic goal.

Your strategic goal is for your own internal use, so do not worry about how well it reads. All that matters is that it captures how you want your life to be and that all the words mean the same thing to each person. Once you have a rough draft of the quality of life statement you are ready to describe the things you have to produce to create it.

Step 2: Identify Forms of Production

Meeting Quality of Life Needs: Ask: "What don't we have now or what aren't we doing now that is preventing us from achieving this?" Re-phrase the answer in positive terms and you will know what you have to produce. For example "financial security" could be achieved by producing a "retirement plan" or an "estate plan."

If being "debt free" is one of your desires, then profit will need to be a form of production. Specify the sources of profit in general terms. For example, profit from crops but do not specify the kind(s) of crops because markets and other conditions change. And, these will be the kinds of decisions that will need testing against the strategic goal.

Meeting your Stated Purpose: If your organization was formed for a specific purpose, here is where you state what you must produce to meet that purpose.

For example, if you are engaged in rangeland restoration, you must produce a collaborative network of land and livestock owners who are knowledgeable of the methods you use.

Creating your List: Avoid common errors:

1. Are all the ideas contained in your quality of life statement covered?
2. Have you included what you must produce?
3. Only list *what* you need to produce, not *how*. *How* is a decision that needs testing.
4. Did any conflicts arise? If so, you were probably discussing *how*.

These first two parts of the strategic goal primarily address immediate needs. The final part looks to the future.

Step 3: Describe the Future Resource Base

Describe the resource base that you will need many years from now to sustain the production needed to maintain the quality of life you want.

If land is a part of your goal, think of the resource base over the very long term. Describe how it will have to be 1,000 years from now.

Several elements are involved: the people, the land, the community you live and work in and the services available in that community.

The People: Most organizations have customers, clients, advisors, suppliers, etc that are critical to their future. You have already listed many of these when you defined your resource base. These are people who make no management decisions but can influence them.

We cannot describe these people far into the future. But, we can describe how we must be far into the future. So, describe how you must be seen for these people to remain loyal to you.

In the case of clients, for example: describe yourself as you would have to be in the future—honest, reliable, prompt, professional, etc.

The Land: Even if what you produce does not come directly from the land, it needs to be included because, in the long run, the well being of any entity (family, business or community) depends on the stability and productivity of the surrounding land.

When what you must produce comes directly from the land, you will need a detailed description of what the land will look like and how the ecological processes (mineral and water cycles, energy flow, community dynamics) must be functioning far into the future.

The task is to describe the land in terms of how the four ecological processes will have to be functioning in order to sustain what you have to produce to support the quality of life you want over many future generations.

When dealing with several environments (rangelands, crop lands, forests, etc), you should create separate descriptions for each of them. If extensive land areas are involved, it will be helpful to map these descriptions.

As with the rest of your strategic goal, the land will be redefined over time as you learn more about its capabilities and testing decisions force you to be more specific.

The Community Your Live or Work In: Sometimes you might want to describe the type of community you want to live in since many of your future resources may be derived from that community and what you produce may depend on it. Then you can begin making decisions that take you toward that.

The Services Available in Your Community: Related to the community are the services that you need to be available in your community. Things like a library or the medical services are generally more a concern in rural communities.

Once you have thought through and described all the elements that make up your future resource base, you will have a strategic goal.

The development of a strategic goal is a lengthy and difficult process. That is why you should start with a temporary strategic goal and modify it as you go until it becomes the perfect fit.

Refining Your Strategic Goal

It takes time for people to express more than superficially what they want in terms of quality of life. So, forming a strategic goal can take several years. But few want to wait that long to begin putting their situation right. That can be overcome by first forming a temporary strategic goal and starting toward that.

Form your temporary strategic goal quickly (in hours rather than days) and then start to use it to make real decisions right away (even on the same day) so people can begin to see the value of it.

Temporary means the goal is open to discussion and improvement. Until it expresses what people genuinely desire, they will tend to go back to arguing about tools and actions. Furthermore, they are more likely to find strategic decision making "too difficult" or "too much trouble" and to be tempted to return to the old familiar ways. But over time, the strategic goal will come to reflect what they truly want and the rest becomes relatively easy.

The temporary goal will have a great deal in common with the more permanent strategic goal that develops from it. The main difference will be in the degree of commitment people have to them. You cannot force commitment. If you try, people will only pay it "lip service" and it will lose its power.

Common Mistakes: Mistakes are common, if not universal. One of the first is to begin from the standpoint of today's problems. This leaves no incentive to go beyond the problem to what people really want.

Some are tempted to start with describing the future resource base in the belief that they are already in too much conflict to be able to start anywhere else. But this usually exacerbates the conflict and ends in deadlock.

Across all cultures there is a great deal of similarity in the quality of life desired. So, always begin with an expression of the quality of life you desire.

Do's and don'ts.

- Make your strategic goal 100% *what* you want and 0% *how* you plan to get it.
- Do not allow any prejudices against future tools or actions to appear in the goal.
- Do not even try to prioritize ideas expressed in the quality of life statement.
- Do not quantify the forms of production in any way. Nothing requiring testing should be in the strategic goal.
- The strategic goal should be formed by the people who will be living it (those directly involved in management), not outsiders.
- Keep in mind that levels of commitment will exist when some of those who formed the goal are removed from day-to-day operations—a common problem with boards of directors and absentee owners.

Common Challenges: A primary problem is the tendency to get so specific that the strategic goal takes pages to express.

Two of the greatest challenges are: 1) reluctance or inability of people to express what they value most and 2) a group that is already not functioning well.

Expressing Personal Values: Most of us want to live our lives in a way that is pleasing, rewarding and based on the things we value most. Making a decision that is not in line with our values is illogical.

Many people find it hard to talk about what they value most let alone write it down. Regardless, the first step is to identify what is most important to you. The rest of the strategic goal builds from that.

Some businesses develop mission statements that are, all too often, written by the leader or a committee and imposed on everyone else. Often these are superficial, politically correct statements strictly for public consumption. By contrast, the wording of your strategic goal will have meaning for all who produce it and are going to work and/or live by it.

Dealing with Dysfunction: There are no easy answers for groups that are not functioning well. It only takes one person suffering from low self-esteem, mistrust, etc to make the whole group dysfunctional. There are two different ideas on how to deal with such situations.

One is to work a lot on team building, trust building, conflict resolution, personal growth and so on. The down side of this is that it may be years before you have a well functioning group.

The other approach is to form the strategic goal first and acknowledge up front that people in the group are not comfortable with one another. Then address the problems as they emerge. In very large companies you generally have no other option (provided you can get them to stop fighting long enough to form a temporary strategic goal beginning with how they want their lives to be).

A facilitator might be able to solve that by informing them that they will likely all be dead within the next 20 years. Then ask them to think selfishly—what would they want the world to be like if they came back in 100 years. Give each of them a piece of paper and send them off to write down their thoughts. Take the papers when they are done and write the comments on a blackboard for all to see. It is amazing how most (if not all) of them want essentially the same things. Now you can create a quality of life statement and the rest of the strategic goal without a murmur.

Wholes within Wholes: In large, diverse organizations the task is to define wholes within wholes. Make sure the wholes remain cohesive and committed to the well-being of each other, their goals do not conflict, and lines of communication between them remain open. To do this, follow these guidelines:

1. Make sure some of the decision makers in the greater whole also make decisions in the smaller wholes. This crossover is essential to maintaining the core values that unite the groups and keeping the lines of communication open.
2. Create a statement of purpose for each smaller whole. You form a whole within a whole for a specific purpose which needs to be expressed because each smaller whole supports the greater whole. When the smaller whole defines their strategic goal, one or more of their forms of production will relate specifically to their statement of purpose.
3. Make sure the future resource base described in each entity's strategic goal addresses client-supplier relationships. At one time or another each is a client or supplier of the other. The decision makers in each whole need to describe their behavior as it will have to be to maintain productive relationships.
4. Clarify financial arrangements. The greater whole may make money available to the smaller whole. If it does and expects a return on investment, that needs to be clarified up front. The numerous transactions that may take place among the various wholes will proceed more smoothly if parameters are established from the beginning.

Conclusion: If you form your strategic goal properly, you will see a change for the better even though you might muddle through the rest of our recommended approach to management. On the other hand, muddle through forming your strategic goal (drop a part of it, don't write it down, don't check for agreement) and do the rest to perfection and eventually you will end up in despair. It is that important.

Testing Decisions against the Strategic Goal

Henceforth, all the decisions you make will be tested against your strategic goal.

There are seven simple tests consisting of one or two questions each that you should ask yourself before implementing any decision. The testing should take only a few minutes. Once you internalize the questions, it will be accomplished in seconds. You will begin to do it subconsciously.

Do not dwell on any one test to the extent you lose sight of the picture formed by scanning them all. This approach will increase you confidence that your decision is economically, environmentally and socially sound.

You will make decisions in the same way you always have. But, your purpose is to test them for soundness and to make sure they will take you toward your strategic goal. If a decision fails one or more of the tests, you may decide to abandon it, modify it, or proceed anyway. But at least now the decision is a rational management choice.

The seven tests are summarized below. (The only rule of order is that the Society and Culture one should always be last. Any test that does not apply to a particular situation may be skipped.)

1. **Cause and Effect:** Does this action address the root cause of the problem?
2. **Weak Link:**
    - *Social*. Could this action, due to prevailing attitudes or beliefs, create a weak link in the chain of actions leading toward your strategic goal?
    - *Biological*. Does this action address the weakest point in the lifecycle of this organism?
    - *Financial*. Does this action strengthen the weakest link in the chain of production?
3. **Marginal Reaction:** Which action provides the greatest return, in terms of your strategic goal, for the time and money spent?
4. **Gross Profit Analysis:** Which enterprises contribute the most to covering the overheads of the business?
5. **Energy/Money Source and Use:** Is the energy or money to be used in this action derived from the most appropriate source in terms of your strategic goal? Will the way the energy or money is to be used lead toward your strategic goal?
6. **Sustainability:** If you take this action, will it lead toward or away from the future resource base described in your strategic goal?
7. **Society and Culture:** How do you feel about this action now? Will it lead to the quality of life you desire? Will it adversely affect the lives of others?

Speed is essential to the process. If you cannot quickly answer "yes" or "no" move on to the next test. In most of the cases where you have to bypass a test it is because you do not have enough information to know whether the decision passed or failed.

Don't worry that speed cause an unsatisfactory result. You will be constantly monitoring to insure that it does, in fact, lead you toward your strategic goal. If it does not, you will immediately re-plan and re-test.

If you are trying to alter the ecosystem, you will automatically assume that you are wrong because it is impossible to account for all of nature's complexities. Based on this assumption, you will determine what you should monitor to give you the earliest possible warning of the need to re-plan.

Especially in the beginning, testing will probably make the biggest difference in decisions requiring significant expenditures. It will also make an enormous difference when dealing with crises that suddenly arise. It helps ensure that your decision is in line with your strategic goal and not simply a panic fix.

With time, the process will become so familiar that it will shape your decisions before you make them. You will not consider what to do about a problem until you have identified its cause. You will not do things you do not want to do. In the beginning, testing may show that some of the decisions you have already implemented are unsound. That does not mean that you must immediately discard what you are doing. It does mean that you are now aware of the danger and can plan what to do about it before it is too late. How to regulate the rate of change is, itself, a decision to be tested against your strategic goal.

Power is in the Strategic Goal

Almost all the problems that plague mankind stem from the way we make decisions.

It is no big deal if you don't get it right at first. Everyone has difficulty with the questions in the beginning. Your strategic goal is more important to decision making than the tests will ever be. If you do not have a strategic goal, testing is pointless.

Humans will always bias their decisions in favor of what they really want. So what you really want must be in your strategic goal. Ownership in a strategic goal makes a critical difference. The testing itself will start to show you where your goal needs more clarity.

The testing guidelines are only a mental crutch that helps you see the big picture from many angles simultaneously.

Lesson 3: Four Fundamental Ecosystem Processes: The Focus of Restoration Grazing™

The planet is the only ecosystem that encompasses everything. The boundaries that define any other ecosystem are artificial. Each exists only in dynamic relationship to the others. So, rather than distinguish lesser ecosystems, it is more practical to speak of different *environments* because that does not connote the idea of boundaries to the same extent.

There are four fundamental processes that are common to all environments and the foundation for all human endeavors. Modify any one and you change them all—i.e. one cannot have effective water and mineral cycles and adequate energy flow without living organisms.

These four processes are the focal point of Rangeland Restoration Grazing™. The strategic goal describes the future landscape in terms of these four processes and the subsequent methodology focuses on moving them in the direction of that vision.

Process 1: The Water Cycle

There is a fixed amount of water on the planet that constantly cycles. It falls as rain, hail and snow. Some evaporates. Some runs off into streams, rivers, lakes and eventually to the sea. Some penetrates the soil and sticks to soil particles. The rest goes to underground supplies where it may remain for millennia or find its way back to the surface in springs or through transpiration by deep-rooted plants. Of the water held by soil particles, a small portion is held tightly. The bulk is either attracted to drier particles or drawn away by plant roots and transpired.

The time water spends in the soil is critical to all life, both plant and animal. The water that penetrates the soil is attracted to drier particles. There is no sharp edge between wet and dry soil. Instead there is a gradient from wetter to drier. As water is drawn away from a particle, the particle tightens its hold on whatever remains.

Plants absorb water and the nutrients dissolved in it through their roots. As drying particles yield less and less water, growth slows. However, much can be done to retain moisture and extend the time that plants can grow.

There is little direct surface evaporation in effective water cycles. Any that does run off does so slowly and carries little organic matter or soil with it. There is also a good water-to-air balance which is critical because most plants require oxygen for their roots to grow.

By contrast, with ineffective water cycles much is lost to surface evaporation and runoff. Air and water are not in balance.

In soils referred to as "water logged," water displaces air in the soil due to an impervious layer of sub-soil so that only plants adapted to a lack of oxygen around their roots can grow. A similar effect can occur when the soil has become sealed over with a crust or "cap."

Your strategic goal should describe the water cycle as it will have to be functioning to sustain what you want to produce.

Effective Water Cycles: Rainfall averages often mean little because they seldom occur. Even when annual rainfall is average, the distribution can be very different. However, the water cycle can be evened out by making the rain that does fall more effective.

Effective rainfall soaks in and becomes available to plants and other soil organisms. It replenishes underground supplies and very little evaporates at the surface. At maximum effectiveness, most water either returns to the atmosphere through plants or permeates down into underground supplies.

(Incidentally: it takes roughly 600 tons of water to produce one ton of vegetation.)

Where soils have lost a large proportion of their organic matter, they are unable to absorb much water.

Capping: The nature of the soil is vital to the water cycle. On bare and exposed soil, the direct impact of rain destroys soil structure and frees the organic and lightweight material to wash away leaving the heavier fine particles to settle and seal (cap) the soil surface.

A capped surface not only reduces water penetration but also prevents oxygen from getting into the soil and carbon dioxide from getting out. This affects the organisms responsible for releasing nutrients which leads to nutrient deficiencies. You can see this if you lift the capped layer and inspect it closely. Even very sandy soils will develop a cap.

Soil cover protects the surface from the impact of raindrops which preserves structure and prevents capping. Cover generally comes in two forms: low-growing plants and dead plant material or litter. This not only reduces impact but also slows the flow across the land.

In less brittle environments plants are closely spaced and hold litter in place. Also plant life establishes quickly on exposed surfaces.

But, the more brittle the environment the more opposite this becomes. Bare soil develops easily between plants and over millions of acres. Plants are spaced wider apart which allows wind and water to carry litter away.

Creating an Effective Water Cycle: Proper management can break up the cap, increase organic matter content, improve structure and speed up percolation. A loosened, rough surface or one covered by old, prone plant material will slow the flow and allow more water to soak in.

The most important factor is soil cover followed by organic matter, aeration and drainage. We have available to us management tools that can either promote or destroy cover.

In less brittle environments, no tool causes soil exposure over large areas other than technology. In brittle environments, rest (partial or total), fire (periodic), and (to a lesser degree) overgrazing (grazing misapplied) are the only tools that can expose soil on the massive scale that currently exists.

In non-brittle environments, rest (partial or total) is the main tool for producing soil cover. To increase cover, you should disturb the soil surface as little as possible.

In more brittle environments, the only tool that can provide adequate soil cover over large areas is animal impact. On rangelands, animals can be used to trample down vegetation to provide litter.

On very hard capped soils in the tropics, animals can only break up the capping progressively or when concentrated into very high densities.

Aeration, organic matter and drainage all depend on soil cover. Damage occurs in the more brittle environments from overgrazing or *over-resting* perennial grasses whose roots are the main soil stabilizing force.

Aeration, organic matter and drainage also depend on small animal life forms in the soil. When earthworms are present there is greater water penetration and retention. To encourage earthworms, avoid plowing, pesticides and ensure litter covers the surface.

With an effective water cycle, floods and droughts become fewer and less severe. Floods tend to rise more gradually and subside more slowly. Flood waters tend to be clear.

Droughts are far less severe. More of the previous year's moisture has been stored in the soil and what little that may be received penetrates better. In other words, more water is available for plant growth over a longer period of time.

Furthermore, there is more water available to be released slowly to stream flow, springs and underground aquifers.

As a rancher, what would it be worth to you to double your rainfall? Obviously, that is beyond our control. However, we *can double the effectiveness* of the rainfall you receive which actually has a far greater effect than doubling the amount.

Non-effective Water Cycles: With non-effective water cycles:

- Droughts occur more frequently and are more severe.
- Good plant growth takes place only in short bursts for a few days after the rain.
- Plants don't start growing until later in the growing season.
- Plants stop growing earlier at the end of the growing season.
- Floods are more severe when a high portion of the ground is bare.

Let's put it into mathematical perspective. Bare soil can shed more than half the water falling on it. Thirty inches of rain yields a total of 814,625 gallons per acre. Over a million acres (405,000 hectares), that would amount to 814 billion (with a "B") gallons. If half of that runs off it makes for an amazing flood.

Far too little attention is given to water lost to the high rate of surface evaporation due to exposed soil. In brittle environments, very little attention is given to the amount of soil that is exposed between plants. It adds up to millions of acres. In fact, the brittle environments that cover the majority of the earth's land surface are characterized by non-effective water cycles. Some of the world's largest cities are running out of water. One even tried to increase the amount of water flowing into its storage reservoir by encouraging overgrazing on the catchment area. It worked too. They collected lots of water—and silt too. Dam sites can only be used once. So, before we use them, we need to first be sure the water cycle in the catchment area is effective.

Consequences of a Non-effective Water Cycle: To summarize:

- Increased runoff. Water runs faster and carries more silt. Increasingly frequent and severe floods.
- Decreased water penetration and increased losses through evaporation increase the frequency and severity of droughts.
- Less production in all years. Greater instability and fluctuation in the volume of forage production.
- Falling ground-water tables, drying up springs and wells.

- Unstable rivers prone to flash flooding and intermittent flow.
- Silted dams and eroding catchments.
- Detrimental effects on the other ecosystem processes.

No extensive and extravagant engineering procedure will ever achieve what simply putting the water cycle right does at a fraction of the cost.

How do you Recognize a Non-effective Water Cycle? The signs are basically identical across the brittleness scale but are more easily seen toward the very brittle end.

Look at the soil surface between the plants at their bases. Bare soil means rainfall is less effective. Any increase in bare soil provides a warning.

Other signs:

- Litter banks caught against vegetation.
- Signs of water flow—rills, exposed grass roots, silt deposits, etc.
- Rivers that once flowed all year now only flow in periodic flood stage.
- Water levels lowering in wells; springs drying up.

The evidence of poor water cycles is so obvious in the more brittle environments that most people assume it is natural. Case in point: A Zimbabwe rancher received the "Best Managed Ranch of the Year" award by his conventional range management association. Transects ran on his ranch in an area supposedly maintained by periodic burning revealed that 97% of the soil was bare and eroding. Overall, 95% of this "perfectly managed" ranch was bare ground.

As typified by the western united States, vast areas in the brittle environments are either rested or managed under prolonged grazing with small numbers of animals scattered thinly across the land. When viewed panoramically, much of these areas appear to be vast seas of grass and are conventionally classified as being in "good" condition. However, when viewed more closely, we find that more than 50% of the soil is bare.

Water Cycles in Cities: In addition to the land surrounding the city, urban residents must also concern themselves with the rain that falls within the city. Looking down on most cities, about all a rain drop has to land on are impervious roofs, parking lots and roads. As a result, runoff is extremely high. Very little water is used where it falls. Most of it is channeled into storm drains which empty into lakes, streams or onto wetlands. Much of this runoff contains toxic substances. Runoff from urban areas can destabilize river banks and add to the siltation and flood damage downstream.

In the meantime, city planners try to tap underground sources or transport water from lakes and rivers with little consideration for the long-term consequences or costs (as long as the hapless taxpayer remains willing to sacrifice the fruits of his labor to the folly).

The possibilities for solutions (and associated economic opportunities) are limitless for bright and creative entrepreneurs. Obviously roofs could be used to catch water and store it in cisterns for use in homes and offices. There are paving materials that allow penetration of water. Porous concrete blocks honeycombed with gravel can provide permeable, non-slip surfaces for streets and walkways. The limit is only the imagination.

Conclusions: If your community is prone to flooding or water shortages, the water cycle is probably not functioning well on the surrounding uplands.

If you are managing land, you need to describe what that land looks like when the water cycle is effective. This should include separate descriptions for the different land types under your management—croplands, woodlands, riparian areas, rangelands, etc.

If you want to maintain a wetland environment, you will describe a non-effective water cycle where the soil remains waterlogged.

In other words, the only thing that we absolutely *must* do is stabilize the surface of the soil. Beyond that, everything else is a rational management choice. And, there are tools available that allow us to achieve that choice.

Process 2: Community Dynamics

Beginning with the point that living organisms establish on bare soil, things are never the same again. Once any community reaches the highest level of development possible for its environment, it may appear to be stable. But, it is not. There will be constantly changing patterns taking place within the mature community—i.e. species composition, age structure and numerous other factors are in a constant state of flux. We refer to this process as *community dynamics*.

Of the four ecosystem processes, this one is the most vital. Water and minerals cannot cycle and energy cannot flow effectively unless plants first convert sunlight into energy useable for life and soil cover.

We still have much to learn about the dynamic relationships amongst living things. However, a few basic principles have emerged. The following is not a comprehensive list but they bear most directly on day-to-day management choices.

There Are No Hardy Species: If "hardy" means that the species can endure very adverse conditions, then there are very few hardy species. The so-call "hardy" plants that invade bare ground do so because that is the environment that suits them.

Furthermore, when an organism establishes in a community, it will inevitably alter the surrounding microenvironment. The changes that these new species bring to the micro-environment influence the types of other plants, animals, birds, insects, etc. that find this habitat favorable. Hence, these species increase also. So over time, some species increase and others decrease as the composition of the community makes it more or less favorable for them.

Nonnative Species Have Their Place: Sometimes communities can be catastrophically altered in a short time by the introduction of a new species. The species with no defenses against the introduced species quickly die out. Then, the species that depend on those die out too.

Not all introductions have been catastrophic. An example is the honey bee that was introduced into Europe from North America almost 400 years ago. Introduced plant species have followed the same pattern. We often refer to them as "weeds." But, many introductions have been so successful that we no longer think of them as non-native: A shining example is the introduction of corn from South America into Africa where it is now a staple.

Today, in the united States and spreading rapidly, there is an unhealthy fixation on non-native species. The term *non-native* is purely bureaucratic. Those that use it assume that a species that arrived after a specified date are not "natural" and should be destroyed at all cost.

The nonsense even extends to arguments over whether or not a particular species is "native" to a particular region of the same country. For example, Texas Parks & Wildlife are currently trying to eradicate elk on properties under their control in the mountainous areas of Trans-Pecos Texas. On the other hand, elk are accorded legal status in other areas of the country for no other reason than because they arrived on the continent at roughly the same time as the first humans. And so they are "Native" Americans.

In the same way, the horses' ancestors were present prior to humans but they are considered "illegal immigrants." Just like a legal immigrant, once an "illegal immigrant" arrives it will fill any vacuum that mismanagement provides.

There have been laws passed in some of the states that require land owners to "control" particular non-native immigrants even when they generally play nothing but a beneficial role.

None of this is meant to encourage the introduction of new species. However, once a species has established itself, we are better off managing for the health of the whole community.

So, the landscape you describe in your strategic goal may include species you once considered pests.

Collaboration is more Apparent than Competition: There is far more collaboration than competition in nature. Scientists call this mutually beneficial relationship among species *symbiosis*.

Due to an oversimplification of Darwin's ideas and despite evidence to the contrary, educators push competition as the driving force behind nature. The fact is, over time, new species develop from existing species in an attempt to *avoid* competing for the same ecological niche. Actually, there is no (or very little) *clear* evidence of competition in nature. When problem plants "outcompete" and "take over," it is usually because mismanagement has created the ideal conditions for their establishment. Management creates bare ground (unoccupied by any plant). The problem plant establishes itself and is the only plant present. Automatically, the conclusion is that it *caused* the bare ground.

Wildlife are said to compete with livestock for forage. But, when we properly manage the land there is abundant feed for both.

When you learn to think in terms of functioning wholes, collaboration and synergy, you interpret events differently.

Stability Tends to Increase with Increasing Complexity: Communities in the early stages of succession or that have lost biodiversity are prone to major fluctuations in the numbers and composition of species. Disease and outbreaks of weeds, insects, birds or rodents are more frequent.

Since instability often correlates with weather patterns, we inevitably blame it on the weather rather than a loss of biodiversity as the result of miss-management.

The more complex and diverse communities become, the more stable they tend to be. As the number of species increases, so does the web of interdependencies among them. There is an exception in true deserts where communities are simple but, because the weather is consistently dry, they appear stable.

Most of Nature's Wholes Function at the Community Level: A population of any one species does not constitute an ecologically functioning whole. Furthermore, when we focus on rare and endangered or preferred species, we lose sight of the fundamental importance of the whole community. Some rangelands are classified as being in good condition if the "right" (useful to humans) species are present while, in reality, they belong at the other end of the spectrum because so many species and so much biomass have been lost.

Efforts are made to save rare and endangered species through draconian laws that pay little heed to the communities that support these species.

Biological communities include all living organisms—from the most simple virus or unicellular organisms to elephants. This includes the microscopic life in the soil.

By decimating numerous herding species and their predators, we have greatly reduced the range's ability to trap carbon in plants and soils. Furthermore, the attempt to replace the animals with fire is the most profound environmental error ever made.

Most Biological Activity Occurs Underground: Any changes above ground are likely to cause even greater changes underground. This is because there is more life underground—literally tons of microorganisms and plant roots.

Scientists estimate that 75-80% of the prairie's biomass is underground. Although 60 million bison and millions of elk, deer and pronghorn once inhabited the prairie, underground organisms would likely have outweighed them.

So much life exists below the soil surface that any action that alters the underground community will inevitably change what is above ground.

Change Generally Occurs in Successional Stages: *Succession* is a relatively orderly process of change. Visualize the process on a tropical island lava flow. Shortly after the lava cools but before soil can form, only a few species will be able to establish—algae, lichens and minute organisms.

However, as soon as they do, the microenvironment becomes different and the creation of soil begins.

Gradually other organisms join the community as the environment begins to favor them. They further change the microenvironment and succession accelerates. Complexity, productivity and stability increase and the microenvironment continues to change until something limits the successional process. This is typically climate or some other obstruction to further soil formation.

Whether the outcome is jungle, desert, savanna, or whatever else, the community is always dynamic. A particular species will appear and its population will build up as its requirements for establishment are met. At the same time, as the community advances, a population may find its requirements for reproduction are no longer ideal. It will decline in numbers and may even disappear as the successional process advances.

Defying Succession: Succession is never a smooth or straightforward progression from simplicity to complexity. Some species seem to try to maintain their own ideal environment. Others develop adaptations that limit the success of species that graze, browse or prey on them. For example, prairie dogs create open country around their towns to make their predators more visible.

Algae, lichens and mosses are often the last plant life left in a deteriorating biological community in the more brittle environments when thousands of years ago, they were probably the first life to establish. Once the soil surface becomes encrusted, succession stagnates. But, once the crust is broken, succession advances again.

Succession and the Brittleness Scale: In brittle environments with exposed soil surfaces and daily and seasonal extremes, succession starts with the greatest difficulty.

The process starts more easily on soil covered by old material and on ground cracked by weathering or broken by the physical impact of animals or machinery. There are two exceptions.

Although we really do not understand why, if fallen material all lies in one direction it suppresses pant growth. On rangelands snow and wind lay moribund bunch grass in one direction. Hail and animal impact scatter it. In forests pine needles may suppress plant growth. But new plants will establish with disturbance from animal impact.

The other exception is areas where soils become puffy and soft from alternate freezing and thawing. These soils have broken and rough surfaces, yet secession does not progress due to the lack of soil compaction.

This highlights the need to view the community as a whole. Managing plants or animals in isolation is meaningless and likely to be damaging. Any brittle environment lacking large grazing animals is unlikely to develop its optimum level of complexity and stability. The presence of grazing animals is necessary on brittle environments.

By contrast, in non-brittle environments succession starts with ease from any bare surface without the aid of some physical disturbance. In fact, it is hard to stop the rise of succession. A jungle badly damaged by being cleared for farming will return to full complexity. Although complete recovery may take decades, the advance from bare to covered soil will be very fast. In other words, non-brittle environments are highly resilient to drastic disturbances.

Community Dynamics and Management: If you want to produce livestock or game, you probably want productive grassland. In one case that could mean advancing succession from desert shrub while in another it may mean preventing pastures from returning to forest.

If you wish to favor a species, you must direct succession toward the environment that is optimum for that species. Although it may be a necessary interim step, simply protecting the species will not save it.

If you start with an environment that contains undesirable species, your strategic goal should describe a community that is less than ideal for that species and more suited to what you want to produce. To do this, you will need to know the basic biology of the species—what specific conditions does it need to survive at its weakest point. A little effort at that point will influence whether it decreases or increases.

Do not fall into the trap of seeking a monoculture when you are trying to increase or decrease a species. By nature, succession moves toward greater stability and complexity.

Historically, we have ignored succession and dealt only with its effects in our attempts to eradicate a pest species.

Some fluctuation of species is natural (especially) among short-lived organisms with high reproductive rates that characterize lower successional communities. However, prolonged downward movement to lower levels of succession is unnatural and almost always betrays human intervention.

Leaving it to Nature: There are those who advocate leaving the management of the environment to nature based on the assumption that nature knows best. But, lands that were abandoned and left to nature centuries ago are still deteriorating. They will never recover (at least on a human time scale) unless we use animals to simulate the effects of large herds and predators that made them function long ago.

Conclusion: You should describe what is needed in your strategic goal—i.e. communities that are rich in plant and animal species or *biodiversity*.

If you manage land, you will need to be more specific in describing the future landscape.

It is possible that, under special circumstances (like maintaining a particular species), you will need to describe a community that is less diverse—i.e. one that thrives on bare soil.

Process 3: The Mineral Cycle

Like water, minerals and other nutrients cycle. Management can drastically alter the speed, efficiency and complexity of their cycle.

A good mineral cycle implies a biologically active living soil with good aeration and energy underground to sustain an abundance of organisms that are in contact with nitrogen, oxygen and carbon dioxide from the atmosphere. A good mineral cycle provides a wide range of nutrients that are constantly cycling.

The goal is usually to keep nutrients from escaping the cycle and to increase the volume of those cycling in the soil layers that sustain plants. To do this, nutrients must continually be pumped up to the surface by plants.

It is important to remember that the mineral cycle is not independent of the other three ecosystem processes.

Minerals to the Surface: Plant roots are the main agents for lifting mineral nutrients. So you need healthy root systems that can probe into the lower layers of the soil. You also need a wide range of plant species-- some with surface roots and others that are deep rooted. Where shallow-rooted plants (such as grass) dominate, you still need some deep rooted plants. Also, small animals (earthworms, termites and other insects) play a role.

Above Ground to Surface: Plant material returns nutrients to the surface in the form of dead leaves and other residues. Feeding and other animal activities speed up the process.

Nutrients have to move underground in order to be used. The dead material is broken down by the weather, trampling, consumption or decay. When it is broken down by fire or oxidation, the nutrients are lost to gases and their ash is blown or washed away.

Exposed soil reduces biological activity and biological (as opposed to chemical or physical) activity should play the lead role in the breakdown.

In the less brittle environments extremely active communities of small organisms can break down old plant material without larger animals.

In more brittle areas, large animals are critical because, over the year when 50 to 95% of the above-ground material dies, the microorganisms and insects also die off. Large animals are needed to either trample the material or reduce it through grazing and digestion.

Without large animal impact bare ground increases and soil becomes exposed which, in turn, reduces biological activity. Where plants are spaced widely apart and the soil is exposed, it is hard to hold plant litter in place against wind and water.

Fire drastically alters plant material, creates pollution, exposes soil and converts vital nutrients to harmful gasses.

Physical weathering may play little part in less brittle environments because biological decay proceeds so rapidly.

In brittle environments, dead plant material breaks down slowly through oxidization and weathering in the absence of large animals. Also, weathering occurs from the top down and the dead material does not fall to the soil surface where micro-organisms can break it down. This creates a bottleneck in the cycle as nutrients remain tied up in the dead material. Furthermore, large accumulations suppress plant growth.

Especially in bunch grass species with growing points at ground level, accumulation of dead material can lead to premature death of the plant because sunlight is unable to penetrate the accumulation and reach the plant's growing points.

In brittle environments animal activity speeds the breakdown and cycling. Also, in contrast to fire, it achieves this without exposing soil or polluting the atmosphere.

Surface to Underground: Once broken down, nutrients move underground. The two agents for this are water and animal life. This is why, when managing to enhance the mineral cycle, we apply tools that encourage water penetration and animal activity.

Leaching can carry nutrients on down below the root zone. However, the main factor that impedes leaching is organic matter. Less organic material (dead plants and animals) and biological activity, means greater danger of leaching.

The Importance of the Soil Surface: Just as it is with the water cycle and community dynamics, the key to the health of the mineral cycle lies at the surface of the soil. An exposed, capped surface slows biological breakdown, limits air exchange, reduces oxygen, causes excessive carbon dioxide and inhibits root growth. As aeration decreases, so does life, organic material, soil structure and aeration and this chain reaction ripples through the ecosystem. This is an ever present syndrome in the desertification process.

Conclusion: On land where minerals are cycling rapidly, there is very little gray visible by the end of the non-growing season.

If you are managing land, your description should be detailed and can vary according to land type. You need to be sure that minerals do not remain trapped in dead, oxidizing material. You may want to describe what the land will look like when minerals are cycling rapidly with very little gray being visible by the end of the non-growing season.

Process 4: Energy Flow: The Engine that Pulls the Train

All organisms require energy and depend on green plants to convert radiant energy from the sun into chemical energy.

Energy flow is synonymous with the carbon cycle because the storage of energy in most living organisms involves carbon. But, carbon is constantly moving between earth and the atmosphere while energy from the sun is a one-way flow.

The living world is solar powered. All life depends on the plant's ability to convert sunlight energy into edible form through photosynthesis.

Management can drastically affect how much sunlight is captured.

The Energy Pyramid: The flow of sunlight to food for life is represented as an energy pyramid. Some sunlight energy is immediately reflected back into space. Some is absorbed as heat to be radiated back later. A small portion is converted by green plants into food. Thus, green plants form the base of the energy pyramid (Level 1).

About 90% of the energy is lost as heat as you move from one level to the next. Level 2 is the energy stored by animals that eat the plants.

Level 3 consists of predators that eat the eaters of level 1 (including humans).

Level 4 also includes humans and other predators that dine on fish and other predators that fed on level 2.

At Level 5 scavengers and decay organisms reduce stored energy further. Then another level or two of decay organisms convert the remaining useful energy into heat.

At all levels a portion passes straight to decay levels through feces or urine. Thus the pyramid is not exact or tidy. However, the concept of ever-decreasing volume of usable energy holds. None of the energy is actually destroyed or used up. Its form simply changes to heat.

The energy pyramid also extends below ground and affects the health of the other three ecosystem processes. All three require a biologically active soil which, in turn, requires solar energy to be conveyed underground by plant roots and other organisms.

The Energy Tetrahedron: Clearly the broader the base of the energy pyramid, the larger the whole structure and the more energy there is available at every level. We have attempted to broaden this base through extensive and intensive use of fossil fuel based technologies—fertilizers, herbicides, pesticides, and other chemical and mechanical technologies. In the case of low productive rangelands, these technical solutions cost more than they return.

The energy pyramid is properly viewed as multi-dimensional—two tetrahedrons, one above the surface and one below—joined at their bases. The tetrahedron has three sides—time, density and area.

Management can increase energy flow (capture more sunlight energy) at level 1 (the surface of the soil) by increasing the *density* of the vegetation, the *time* during which it can grow and the leaf *area* of each plant.

To illustrate the general management concept, consider a square yard of your range. With your hands, feet or any handy small tool, manually break up the cap on the soil, push down the dead and oxidizing grass and work it in. Now ask yourself, would this prepared square yard grow an *extra* ounce of grass if we only get a one inch rain? Of course it would! Multiply that one ounce by the number of square yards in your ranch and convert your answer into pounds. For a ranch of any size at all, this will amount to millions of pounds. Surprised?

You can do that and more at no extra cost by simply grouping all of your herds into one and concentrating it in a small area to break the capping and trample the old grass down. Then reduce the time the animals remain on any one area in order for the plants to have adequate recovery time.

Time (Duration and Rate of Growth): We can increase plant growth in two ways—by lengthening the growing season (the longer the growing season the more total growth) and/or increasing the rate at which the plants grow within a given time (the less taken from a plant during active growth, the faster it re-grows because there is more leaf area to convert sunlight). Improved mineral and water cycles and increased bio-diversity extend both.

Plant density shows a high positive correlation with periodic animal disturbance. Where animal impact is highest, plant density will be higher and *vice versa*. And, the more brittle the environment, the more plant density is affected by large animal disturbance. Using fire and animals, animals alone or machines designed to imitate animal disturbance can all lead to closer plant spacing. However, when any of these are miss-applied, they can all lead to an increase in bare ground. Fire used alone always does that. Whereas, using animals alone can actually make money, anytime you add either fire or machines, you incur unnecessary costs.

Leaf Area: A very dense stand of narrow-leafed plants will trap less sunlight than a moderately dense stand of broader-leafed plants. Thus, this implies the desirability of increasing the proportion of broader-leafed plants.

Plants adapt to three basic types of growing conditions. There are: wet-environment, middle-environment and dry-environment plants. The wet and dry type plants resemble each other in that they both have impervious skins that result in little water being passed through their systems. They both have narrow leaves which reduce the area exposed to sunlight and as a result reduce transpiration.

Middle type plants have open, flat, and broad leaves, no thick protective skins or mechanisms to shut off pores. They grow rapidly when conditions are favorable. Middle type perennial grasses cure to red or gold in the dormant season which indicates the forage is more nutritious.

To increase energy flow, the plant community needs to be shifted to middle plant types that spread their broad leaves to the sun and grow rapidly.

In addition to causing grass plants to grow closer together, animal impact and severe grazing (without over-grazing) cause many species to produce more leaves and less fiber. This, in turn, increases the digestibility and therefore the energy available to animals.

Using Technology to Increase Energy Flow: Although it has historically been the cornerstone of American agriculture, attempting to increase energy flow by using fossil fuel intensive technology is very rapidly becoming economically unviable. Often times, when the energy budget is carefully monitored, we find that more energy is consumed in the process than is provided by the result. This economic reality is especially troublesome on rangelands that are inherently of low productive value.

The fact that traditional, energy expensive brush control and re-seeding projects do not pay is readily observable when driving the roads of the rural Southwest. Although one sees where some of this has been (or is being) done, the telling indicator is "where"—almost exclusively on lands belonging to owners with substantial income from sources other than ranching. The families that still depend on ranching for their livelihood cannot afford the folly simply because it does not pay—costs exceed benefits even after government subsidies.

On the other hand, benefits exceed costs for the more affluent (often times absentee) ranch owner once government subsidies and tax advantages are factored into the analysis.

Once again, we see how interventionism distorts the free market allocation of scarce resources and unnecessarily squanders our natural resource base.

Conclusion: To describe (as you might have in your strategic goal) what the land might look like if energy flow is high: the soil is covered by vegetation, there is a great variety of plants and they stay green and growing longer and wildlife is plentiful.

On most rangelands, energy flow can be increased by manipulating grazing and animal impact to achieve maximum growing time, plant diversity and leaf area. The amount of energy (hay) you have to buy to supplement what your own land does not provide is a measure of success or failure.

In sum, if water and mineral cycles are effective and biodiversity is high, then energy flow will be maximized.

Lesson 4: Managing the Ecosystem

Introduction

The tools we have for managing our natural resources include everything that allows us to alter the environment—i.e. everything from stone spears to genetic engineering.

No tool is good or bad. Only when all the factors bearing on a situation are known and the strategic goal is taken into consideration, should any tool be judged suitable or unsuitable at a given time.

Tool 1: Money and Labor

One or both of these tools is always required. People once supported themselves by applying creativity and labor directly to raw resources. During this time, tokens represented actual wealth, goods or services. However, this ended once there became only one source for the creation of "money" (government) which allowed more to be put into circulation than the actual goods or services actually exchanged. Thus we had the birth of the insidious hidden tax that we now call "inflation."

This centralized creation of money has seriously destabilized the relationship between money and the wealth it represents. Your strategic goal should include a sustainable source of wealth that is more vital than the traditional symbol (the currency that is currently in circulation). Three most basic sources of wealth represented by the dollar is what Alan Savory calls: mineral dollars, paper dollars, and solar dollars.

Mineral Dollars: These are represented by the money derived from human creativity, labor and raw resources such soil, timber, water, oil, coal, gas, gold and other metals. Mineral wealth has two primary characteristics:

First, it can be used cyclically over time—i.e. recycled. Or second, it can be used non-cyclically (consumed) as is the case of oil and gas as well as the soils involved in mainstream agriculture.

Paper Dollars: Many of us acquire money through labor and creativity alone and do not directly consume other resources. Paper dollars are backed only by confidence in government and the banking system. When that confidence is lost, paper dollars can completely lose their value over night.

Solar Dollars: We can generate wealth from human creativity and labor applied to sources of energy such as wind, falling water and many others but most of all the sun. Thus, solar dollars are non-cyclical but inexhaustible. This category of wealth is the only kind that actually feeds people.

Conclusion: For perspective: Some 24 billion tons of soil erodes each year from the world's agricultural lands. This is enough to fill a train of freight cars that would stretch from the earth to the moon *five times*.

When testing your actions and decisions against your strategic goal, keep in mind that only solar dollars combined with mineral dollars will allow you to produce the biological capital that will sustain your efforts.

Tool 2: Human Creativity

No two ranches are alike. Each is as unique as your fingerprint. Therefore, every situation requires management that must be an original product of human imagination, and even that must evolve as the situation changes. Creativity is critical and needed constantly.

Consultants often play a role like an extension agent—telling clients what to do. This approach always produces imperfect advice because the consultant is an outsider looking in. It is better to give the client the knowledge and then allow them to do the thinking for themselves. The consultant's role is confined to bringing in outside experience to stimulate the client's thinking.

Because the consultant is on the outside looking in means that he/she has a higher than average chance of being wrong. Consequently, we only work as consultants with people who are willing to learn how to make decisions and plan for themselves. This is the only way possible to proceed from the client's point of view and not ours.

Fortunately, creativity is not strictly genetic. It depends on mental, emotional, and physical health, the environment and the commitment the person has to achieving their goal.

The relationships between the people involved has a bearing on the creativity of each of them and the group as a whole. The person at the top sets the tone. It is his/her beliefs and behavior that have the greatest impact on the creativity of the group.

The manager's most vital responsibility is to create an environment that nurtures creativity. The creativity of the group is greatest when the leader's actions display trust and confidence, the work is meaningful, and all are free to express their ideas and feel valued. Such an atmosphere cannot be faked.

Subconscious worries and stresses sap creative energy. Many of these stresses are associated with crisis management which is associated with the way we make decisions. Also, poor time management leads to crisis management and stifles creativity. Fortunately, these are entirely within our control.

Whatever time management system used, the keys are habit and trust. All your ideas and commitments should immediately be recorded in one place as a matter of habit. Once the habit is formed you begin to trust the procedure and let go of your subconscious worries. This will free up the mind for creative thought.

Conclusions: Every person in the world has the same amount of time. How time is managed makes a difference in the quality of our lives and in what we are able to create. In turn, creativity is the key to using money, labor and other tools successfully. It is the only tool that can produce a strategic goal and plan for its achievement.

Tool 3. Fire

Be cautious of the argument that fire can be used as a management tool because it is "natural." Although it has existed for millions of years, its use by humans for modifying the ecosystem is relatively recent. Since its use began, the frequency of fire has increased geometrically as we have used it with great abandon.

Natural fires (caused by lightning, etc) occur only very rarely in comparison to man-made fires. A booming population, matches and government agencies have radically increased its use in modern times.

The increased incidence of fire and a reduction in animal disturbance due to dwindling numbers are the prime forces of desertification. Evidence of this is the increased flow of water over Victoria Falls in Africa since 1948 due to the increase in the use of fire on the catchment areas above the falls and the continued decrease of the big game herds.

The earliest people of North America used fire a great deal and significantly altered the landscape. Where animals (bison) did survive they mitigated the damage done by fire. In fact, it was the combination of fire, grazing and predator-induced animal impact that produced the lush grasslands found by early Europeans, not fire alone.

As with all of the tools, the decision to use fire should only be made within the context of your strategic goal as it relates to the four ecosystem processes.

Effects of Fire on Biological Communities: Deciding whether or not to use fire requires an understanding of what it does and does not do to the biological community.

Soil Surface: Fire exposes the surface of the soil which is central to the management of all of the ecosystem processes. Fire has its most lasting impact where soil cover takes longest to form—i.e. in brittle low rainfall environments. Low rainfall produces less vegetation and bare soil makes the rain less effective. As a result, the effects of fire can persist for years. Then, if the fire is followed by total or partial rest (low animal impact) as it is on most prescribed burns, soil cover is even slower to accumulate.

Plants: Some sensitive perennial grasses disappear if burned. The majority thrive as burning removes all of the old material that kills grasses when it accumulates. Most of the trees and shrubs that are considered problem species are resilient when burned. But, although they appear dead immediately after the fire, they soon re-sprout more stems than before.

Fire not followed by any other soil disturbance causes major changes within a community. As with any influence that creates the same microenvironment over large areas, it favors the few species adapted to it. A uniform microenvironment leads to fewer species and a near-monoculture of low stability.

Managers can choose either a hot or a cool burn depending on the effect desired. Hot fires involve a lot of dry material that burns fiercely with large flames. Limited dry material produces a "cool" burn with small flames.

It takes a very hot fire to kill shrubs and trees. The problem is that a fire that hot will also kill perennial grasses. So, if you absolutely must burn a grass-land that you are trying to restore or maintain, always burn with the idea in mind of "invigorating grass" and never burn with the idea in mind of "killing brush."

Animals: Like plants, animal response to fire varies greatly. Some are attracted to fires by the easy pickings of food from fleeing insects. If the larger game animals are left alone, they will usually just calmly get out of the way. Some animals seek out burned areas soon after the fire, especially when the first green re-growth appears.

We have made the mistake of noting only the immediate impact of fire on the adult populations of plants and animals. We have not watched what happened to the ecosystem processes in terms of what these things need in order to reproduce over prolonged time. A short-term benefit for an adult population can encourage further burning which may destroy that population in the long run. In fact, both the animals and the plants of grasslands provide many examples of the destruction of too frequent burning.

Fire Alone Does Not Maintain Grasslands: The old myth that fire maintains grassland arose from cursory observation and short-term research. However, maintenance of stable, highly productive grassland is much more complex than that. Time has proven that the grasslands temporarily created by fire masked a serious long-term desertification process. The profound changes were so gradual that they went unnoticed.

Kruger National Park in South Africa is burned every two or three years in an effort to maintain the grasslands that are turning to brush due to partial rest. Managers believe over-browsing and overgrazing are due to too many animals so they cull large numbers every year which, in turn, only increases the over-rested vegetation that is subsequently burned. No soils anywhere can withstand burning this frequently.

Shooting animals and then using fire to replace their role is not confined to Kruger. In Zimbabwe elephants are destroying baobab trees because frequent burning has caused the grasses they favored to be replaced by more fibrous, less nutritious varieties.

Fire and Atmospheric Pollution: In the past most atmospheric pollution was contributed to petrochemicals but biomass burning is also a significant contributor. Every second the emissions from a 1.5 acre vegetation fire is equal to the carbon monoxide emissions of 3,649 cars and the nitrogen oxides produced by 1,260 cars. Today, half (3/4 in Africa) of the worlds 1.85 billion acres of savanna land is deliberately set on fire every year to release 3.7 metric tons of carbon.

The argument that grassland fires are not a pollution problem because plant re-growth absorbs the same amount of CO_2 fail to consider the effects on plant spacing, composition, soil cover, water cycles, etc. which generally leads to less biomass which, in turn, means that less CO_2 is absorbed.

And, as usual, government interventionism exacerbates the problem. If a rancher of so-called "public lands" (65% of the land mass west of the 100th Meridian) determines the need to knock down old dead plant material he will, more likely than not, find that government regulations prohibit running animals at the necessary density. Therefore, his only other choice is fire which does, indeed, remove the old material but at the same time exposes soil and does nothing to break the capped surface while releasing carbon into the atmosphere.

So, when considering the use of fire, keep two things in mind: 1) your strategic goal and 2) the viability of the whole population structure (not just the adult organisms) is critical.

The following is a summary of the effects of fire on the four ecosystem processes at either extreme of the brittleness scale.

Very Brittle Environments

Community Dynamics: Fire exposes soil and new cover develops slowly. It may cause a short term increase in species diversity but repeated fire reduces diversity. It stimulates most woody species and kills only a few. It can produce a mosaic vegetation pattern creating edge effect and a zone of greater bio-diversity. Over time, it will reduce the land's ability to remove CO_2 from the atmosphere.

Water Cycle: Fire reduces its effectiveness by exposing the soil and destroying litter that slows down water flow and maintains structure and aeration.

Mineral Cycle: Fire speeds the cycle in the short term by converting dead material to ash. It also releases pollutants into the air. Because it exposes the soil surface and changes the micro-environment, if used repeatedly it will slow the mineral cycle in the long run.

Energy Flow: Fire may produce an immediate increase in energy flow by removing old material that inhibits growth. However, exposed soil leads to less effective mineral and water cycles and changes the plant community and therefore reduces energy flow in the long run.

Non-brittle Environments

Community Dynamics: Fire has little long term consequences other than for the atmosphere. The higher humidity facilitates a rapid return to complexity. Close plant spacing minimizes soil exposure.

Water Cycle: Fire tends to damage the water cycle by exposing the soil. But the effects are temporary due to better moisture distribution and more rapid succession.

Mineral Cycle: The mineral cycle appears to speed up but this is an illusion. Fire was the main tool in slash and burn agriculture because it freed the nutrients for human use by growing plants in the ash. However, the effect was temporary and such systems broke down rapidly unless the land was rested for 20 years or more between fire use periods.

Energy Flow: In the short run, fire disrupts energy flow but it recovers quickly. Frequent fire damages all four of the ecosystem processes.

This summary assumes that the fire is followed by rest (the common practice) which is not the best thing to do because it is impossible to use fire as a tool without also using one of the other tools.

Tool 4: Rest

As used here, this refers to rest from major physical disturbance of plants and soil by large animals, machinery, fire and other natural means. Withholding all of these for extended periods of time is called *total rest*. *Partial rest* occurs in the presence of large animals scattered so thinly on the land that a large proportion of the plants and soil remain undisturbed.

Brittle and non-brittle environments react in different ways. Prior to the human use of fire and spears, the major grazing areas of the world were seldom, if ever, rested on a large scale. Our use of partial rest in combination with misguided use of fire has had the most devastating effect on brittle environments. Intentionally resting land and justifying it as "leaving things to nature" has changed natural relationships and destroyed our rangelands.

Effects of Rest in Non-brittle Environments: Old plant material breaks down quickly in these environments. Decomposition starts close to the ground where the insect and microorganisms populations are the highest. Thus the dead leaves and stems weaken at the base and fall aside to allow light to reach the growing points. This process affects bunch grasses and woody plants as well which allow communities to remain stable and complex. Water carries away very little debris. Even prolonged rest has little adverse effect on the four ecological processes.

After these environments are reduced to bare ground they respond to rest and quickly return to their former complexity and stability.

Rest does not prematurely kill perennial grasses in these environments. However, it will allow succession to move on to shrubs and eventually to woodland or forests in the absence of some other limiting factor. So, if the strategic goal describes a grassland, rest should be avoided. It is necessary to gaze perennial grasses severely but not over-graze them. In addition, you can create a mosaic of grass and woodland by partially resting certain areas but not others.

It is an error to assume all environments respond the same way to rest. The key to land management is the surface of the soil. In non-brittle environments it is essentially impossible to keep vast areas of the soil bare. The soil surface will quickly become covered over again if rested.

Effects of Rest in Very Brittle Environments: Old plant material breaks down slowly in these environments and succession advances slowly as well. Organisms of decay are present only when moisture is adequate. When rested, old material breaks down through oxidation and physical weathering. Since the tips of leaves and stems are the most exposed to the elements, they break down first. This has little effect on woody plants but severely hinders perennial grasses that have their growth points at ground-level where they are shaded.

Thus, old plant material weakens perennial grasses and will eventually kill them. It also adversely affects grazing animals who are trying to balance their diets by avoiding old oxidizing material.
Some perennial grasses can withstand rest. These usually have their growing points high enough above the ground to allow the plant to stay alive even with un-decayed material choking it. *Tobosa* grass is a good example. Others are of very short stature (*grama*) or have sparse leaves (*aristida*) that allow light to reach ground level growth points.

These few species of perennial grasses dominate the rangelands of the western united States. In contrast, densely populated areas with large numbers of animals (India) are dominated by species tolerant to overgrazing. In both cases, the amount of bare ground between the plants is about the same. Both are using partial rest which promotes bare ground.

When the soil surface stays undisturbed, new plants do not replace the dead. Soil capping is not conducive to seed germination.

Brittle, low rainfall environments subjected to extended rest have wide bare spaces between plants. The remaining plants survive because, although the centers may already be dead, sunlight can reach the growing points around the edges. No seedlings survive on the bare exposed soil, with the exception of a few forbs.

Closely spaced grass plants establish after the soil is disturbed. However, if the land is then rested, the closely spaced plants kill one another as old growth shades even the edges of the neighboring clump. If your strategic goal calls for woody species and you have adequate rainfall to support these species, you would consider continued rest. Dead perennial grass centers are a good micro-environment for seed germination and their dead roots provide an excellent medium for root penetration and seedling establishment. Most of the brush that is a problem today is a result of heavy (mostly partial) rest.

Resting very brittle environments can eventually destroy woody species. Once all grass plants have been oxidized, the bare surface becomes capped. Seedlings can establish only in cracks and growth is difficult unless some form of disturbance is applied.

Effects of Rest in Less Brittle Environments: In environments in the mid-range of the brittleness scale, it is difficult to be sure at what point rest effects change from improving soil cover, energy flow and perennial grass plant health to damaging them. The effects also take longer to show up—it may be years before they become apparent.

These environments lie to the right of the midpoint of the scale—too brittle to sustain grassland without repeated disturbance and too little rainfall for succession to woodland. Totally rested bare soil is the result. Partial rest would only have slowed the process.

About 2/3 of the earth's land is brittle to various degrees. Paradoxically, since the beginning of agriculture, it has been managed in such a way as to produce *both* partial rest and overgrazing.

Some savanna-woodlands and grasslands in high, but seasonal, rainfall areas show signs of brittleness. Old plants oxidize and break down from the tips rather than rotting near the base. Yet, if they are rested long enough, they will pass on to forest and assume the characteristics of less brittle environments. The tree canopy keeps humidity at ground level fairly constant thus keeping microorganism populations constant.

Also, there are places (riparian strips, for example) where rainfall is lower but tap-rooted woody plants provide cover and stability because subterranean moisture allows a dense enough stand. If rested, these will pass on to stable woodlands. It is foolish to assume that riparian areas will remain healthy if rested. Especially those that lack sufficient subterranean moisture to maintain dense, woody cover will seriously deteriorate. Unfortunately (but expectedly) the united States government's policy is to fence off and rest riparian areas.

Delayed Effect of Rest in Brittle Environments: Most governments have long been advising ranchers that range deterioration is due to overgrazing which is, in turn, caused by too many animals and therefore the solution is to drastically reduce animal numbers.

To demonstrate the advantages of destocking (rest) the uS government fenced off demonstration plots. Once protected from overgrazing and unimpeded by old growth they indeed grew lush. But, officials were misled by the power of paradigms and the time delay before the effects of rest became apparent.

I am aware of a situation where a Soil Conservation Service report was written in 1981 but a 1930s photograph was used to illustrate the advantages of rest. Had a 1980s photo been used, it would have been apparent that those plots were deteriorating as badly as any land. Why question what we already know? This is a vivid illustration of why I laugh every time I hear a range manager refer to what they do as a "science."

With fewer animals on the land, partial rest increases, but overgrazing still continues. The range becomes dominated by bare ground, rest-tolerant perennial grasses, brush and weeds. Seeing this decline, animal numbers are cut further and millions of dollars are wasted on expensive mechanical and/or chemical "solutions."

Partial and Total Rest Have Nearly the Same Effect: The plots fenced off in the 1930s provide excellent evidence that partial rest can be nearly as destructive as total rest in the brittle environments. The plots excluded all livestock while, on the outside, livestock numbers were low. Over the decades up to the present, stock numbers have continually been reduced without any real recovery. A comparison of the land inside the exclosure with the land outside reveals very little difference. The land is no better outside the plots having roughly the same amount of bare ground.

Researchers who take it for granted rest is natural have pronounced management outside the plots as being successful. They are unaware of or have ignored the fact that, although the plots got better at first, they subsequently declined.

Rest and Crisis Management: Misunderstanding partial rest leads to crisis management. Environmental organizations acquire damaged land and change their management in a predictable sequence. Initially, plants respond vigorously to rest. Progress under nature appears on target as official measurements record the advance. Gradually, however, the measurements reveal the first signs of adverse change. This is not expected so for a year or two they simply hope the problem will go away. But it doesn't.

So, they conclude that fire is needed since it is "naturally maintained grasslands in the past." The situation predictably worsens as fire use becomes too frequent. Next comes technology—plowing, spraying, reseeding, etc. The problem remains until the managers understand the implications of rest.

When confronted with reintroducing animal impact, the typical response is "no bison ever roamed here" or "cattle are not native." This ignores the fact that many animals, other than just bison maintained grasslands.

So little is known about the number and distribution of these animals, the debates tend to be academic. Some people insist that large herds never occurred in some places in the face of evidence that hunting peoples thrived relatively recently in those places.

If a pocket of land once supported a grassland, then there was, by definition, animal impact of some sort. Furthermore, to rule out domestic livestock as a less natural tool than human made fires and bulldozers is downright silly.

What is "Natural?" Some people close their minds because grassland does exist without any evidence of help from herd animals. If plant spacing is close and age structure good in the absence of disturbance by fire or animals, then that grassland is lower on the brittleness scale. If spacing is wide, age structure poor and reproduction primarily asexual, it is higher on the scale. Under prolonged rest, these brittle rangelands are unproductive because they convert very little sunlight unto useful energy.

Before man controlled fire, domestic livestock and technology, plant and animal life evolved together in a symbiotic relationship. Although we go to great lengths to avoid this conclusion, the extraordinary expansion of deserts in only a few thousand years can only be the work of man.

On the one hand, ancient tree rings are considered evidence of drought. On the other, the fact that a primitive population might have upset the environment (water cycle) through hunting, driving game with fire, setting accidental fires, etc. enough to affect tree growth is never considered.

In Canyonlands National Park in Utah, nylon mesh holds mulch to the ground in an attempt to get grasses growing again. It seems to me that herds of cattle are more "natural" than nylon mesh.

Summary of the Effects of Rest

Non-brittle Environments: Biological communities become diverse and stable with highly effective water and mineral cycles and a high level of energy flow.

Brittle Environments: Biological communities decline into simplicity and instability. Water and mineral cycles become less effective and energy flow declines.

Because of the difference in effect, the condition of rested land indicates the brittleness of the area.

Tool 5: Grazing

Grazing can be used as a management tool by manipulating its intensity and timing. Grazing never occurs in the absence of dunging, urinating, salivating and trampling. However, separating the influences and analyzing animal impact and grazing as separate tools helps in interpreting what we see happening on the land.

Strictly speaking, grazing refers only to the eating of grass while, technically, trees, brush and forbs are *browsed*. The tool we are discussing includes both.

Annual plant populations fluctuate widely from season to season. They are difficult to overgraze because they die once they set seed. By contrast, perennial plant populations fluctuate far less and help hold soil in place year around. The lower the rainfall, the greater the role played by perennials. They are easily overgrazed.

Perennial grasses grow in two forms. In the more brittle areas, they grow upright and in distinct bunches. In the less brittle areas they appear as a sward or mat. New growth can occur either very close to the ground near the base or well above the ground at the end of the stems. The position is a likely indicator of the plant's evolution. Those with growth points above the ground likely evolved under little or no pressure from grazing animals. However, those with growth points close to the ground probably evolved in association with grazing animals that kept the plants clear of old stem so light could reach the growth points. If un-grazed for long periods, these plants die.

Overgrazing: If a large animal bites a healthy grass plant down to an inch above the soil during the growing season, the plant suffers a short-term set back. But, in the long-term it gives the plant a boost because it is less encumbered with old leaf while the growing points remain intact.

If the bite occurs in the dormant season, the plant is not harmed because it has no further use for the material it produced last season. In fact, it will have a growth advantage at the beginning of the next growing season because sunlight will not be inhibited from reaching its growing points.

Damage occurs when the plant is bitten severely during the growing season and then is re-bitten again while trying to recover. This can happen in three ways: 1) when the plant is exposed to the animals for too many days; 2) when animals return too soon; and 3) immediately after breaking dormancy when new growth is being taken from energy stored in the roots.

Grasses transfer energy from leaves and store it in their roots as a reserve to carry the plant through the dormant season and support early growth next year. However, severe defoliation causes root growth to cease because energy will be diverted to the re-growing leaves. If roots are not allowed to re-establish this energy, they will die. So, repeated and frequent severe defoliation causes root mass to decrease and, eventually, the plant will die.

So, overgrazing is grazing that takes place on leaves growing from stored energy which is at the expense of the roots. In short, overgrazing is "grazing of the roots."

Overgrazing reduces plant yield and root volume. There is less feed for animals and less liter for soil cover. The decrease in root mass means less material is available for soil life. The soil compacts and looses structure that is vital to effective water cycles.

Types of Grazers: We split grazers into three categories based on how they graze.

Nibblers nip a leaf here and there and seldom overgraze. They are solitary, small and non-herding. Their populations are self-regulating and they never occur in large numbers.

The second category are broad-mouthed animals that feed by the mouthful—buffalo, horses, cattle. They are gregarious and non-self-regulating and severely defoliate plants.

Somewhere between these two are animals capable of nipping an isolated leaf but normally take several at a time—sheep, goats, pronghorn, deer. They are gregarious and non-self-regulating and have been managed as severe grazers.

The nibblers control their populations by some form of breeding inhibition. The non-self-regulating depend on a high rate of animal loss for their survival. They are predator dependent and herding is their main form of protection.

A wide diversity of animals means more thorough utilization of available feed but does not change the basic dynamics of overgrazing. Conventional range managers try to prevent overgrazing by limiting utilization. They have even gone so far as to specify various levels of utilization for individual species and different types of communities. But, overgrazing continues no matter what the level of utilization.

The Impacts of Grazing on the Four Ecosystem Processes

Non-brittle Environments

Community Dynamics: Although plant spacing is naturally close and soil cover hard to damage, grazing maintains grass root vigor, soil life, and structure. Grasslands normally progress toward woodlands but properly managed grazing can halt it at the grassland level.

Overgrazing will reduce root mass but will not expose soil. It will lead to a solid mat of grass. Species that are sensitive to overgrazing can disappear. It can shift succession toward woody communities.

Water and Mineral Cycles: Grazing will not expose soil but will result in even denser cover and enhance these cycles. Where human intervention has reduced forest or jungle to grassland, these cycles will probably never reach the effectiveness of the deeper-rooted woody community. If continued grazing allows progression to stable woody communities, water and mineral cycles benefit.

Energy Flow: Grazing increases energy flow on natural grasslands. But, if it holds a community back from progressing to woodland or jungle, it keeps energy flow down but available for our purposes. *Overgrazing* reduces energy flow in grassland or pasture. Only where it produces a shift to a woody community does it do otherwise.

Very Brittle Environments

Community dynamics: Grazing maintains grassland and increases their diversity and soil cover and retards shifts toward woody species. It increases organic matter, structure, aeration and biological activity in the soil. In forests, it can prevent the buildup of material and thereby reduce fire hazard while maintaining soil cover and organic matter and preventing the death of the forest. *Overgrazing* reduces litter and soil cover, damages roots and shifts the grassland toward a woody community.

Water and Mineral Cycles: Grazing improves both by promoting a more stable root mass, microorganisms and aeration and producing more plants that will provide more litter. *Overgrazing* adversely effects these cycles by exposing soil and limiting the production of litter. It damages roots, decreases soil structure, increases compaction and reduces porosity, organic matter and soil life.

Energy Flow: Grazing increases energy flow, prevents old oxidizing blockages and promotes root and leaf growth. *Overgrazing* reduces energy flow and plant roots and exposes the surface. When combined with partial rest it can change the community from perennial to annual grassland.

In practice, these general guidelines have to be interpreted in concert with so many other influences that one cannot always predict the final outcome of any particular case. But, the feedback loop based on the constant assumption that decisions affecting the environment are wrong, helps overcome this dilemma.

The Relationship between Grass and Severe Grazers: Perennial grasses and grazing animals have a mutually advantageous relationship. Historically, the accumulation of old growth was taken care of by herding animals.

The severe grazer takes a mouth full and moves on leaving other plants untouched. The grazed plant should benefit but often does not because re-growth is palatable and attracts a second bite a few days later. Thus, the plant gets overgrazed while its neighbors go un-grazed. This accounts for how conventional range management practices produce both overgrazed and over-rested plants in the same area. On any conventionally "well managed" ranch with a so-called "correct" stocking rate, one can find large areas where almost all plants are overgrazed and other areas with a great number of over-rested and dying plants.

Over-rested sites will often shift from grassland to herbaceous or woody communities with considerable bare ground.

Stocking rate has little bearing on what happens to an individual plant and overgrazing only applies to individual plants. Thus, applying the word to a whole area is irrelevant.

Although sites that are deemed "overstocked" do contain a high number of overgrazed plants, most also have significant over-rested plants and bare ground. Even where 100% of the plants are overgrazed, planed grazing can not only stop it but might actually require an increase in animal numbers for the good of the land.

Partial rest leads to the same, or greater, increase in bare ground. The two compound the effects of each other. But, overgrazing always gets the blame because the term is used as a synonym for overstocking.

Before they were domesticated, cattle behaved like bison. They remained close together for fear of predation and moved frequently to new feeding ground as the old became fouled. They would not normally return until the dung had decomposed which was long enough in the growing season for plants to have recovered.

When wild herding animals sense no danger from predators their behavior changes. Overgrazing of plants and over-resting of soils increase as the herd remains spread. Dung and urine are scattered so widely that it no longer inhibits feeding or induces movement. Animals remain on the same ground day after day.

Adaptations to Overgrazing: Wild herds often do still overgraze. That is why certain plants developed a defense against it. In Africa, it is common to see a buffalo herd on the same ground for two or three days before moving off. Then another herd of another species may follow, not bothered by buffalo droppings. Plants severely grazed by one herd often are overgrazed by another but this does little damage as long as animal impact is sufficient to ensure plants get replaced.

Some grass species die out. Others sacrifice the center but continue to hang on around the edges. When many plants are overgrazed and the site also receives high animal impact, the whole community may shift toward a solid mat of runner type grasses.

Perennial grasses are remarkably resilient to overgrazing. The one exception is in Mediterranean climates. However, although we do not fully understand the process, they do return when both overgrazing and partial rest are stopped.

Browsing: The same principles apply. Generally woody plants can withstand complete removal of green leaf as long as they get adequate time to recover. They can also withstand continuous severe browsing as long as sufficient forage remains out of reach of the animals.

Adaptations to Over-browsing: Hedging (the most common adaptation) is where plants take on a clipped garden hedge appearance with sort, tightly spaced stems to protect the leaves crowded amongst them.

Sometimes the larger individuals develop a brows-line below which animals take everything while the higher leaves trap sunlight. Some species develop heavy root systems and straggly above-ground parts. A few grow small matted leaves along the main stems.

Such techniques are of no use to seedlings and over-browsing will eliminate them. Without replacements, the population gradually declines. This is apparent along riparian areas where livestock are in kept in low numbers in the belief that this is what is required to prevent such damage.

Browsing enhances the productivity of many forbs and woody plants that are interdependent with the animals that feed on them. Some woody plants have chemical and physical defenses that protect against browsing.

Over-browsing has no relationship to the number of animals but only to the proportion of leaf removed and the time the plant has to recover.

Wildlife and Overgrazing/Over-browsing: Savory relates the story of a national park they created in Zimbabwe. Their first mistake was removing the people, the primary predator of the game (especially elephants). As a result, the elephants were now able to remain near the river for prolonged periods of time and the destruction of the vegetation quickly grew serious. Their research proved that there were too many animals so they began a culling program. After nearly 20 years of regular culling, the damage to the vegetation remained as bad as ever. The large old trees still produced seed and much of it germinated but none survived beyond the seedling stage. The few remaining elephants continue to over-browse.

The role of humans as a predator had kept the game wild and moving. Although the elephants were culled, they did not know who was doing it so they lost their fear of man. Their response to human scent became very different from what it was when the park was originally opened. The deception was necessary because it is a national park and tourists need tame elephants. So now they linger too often and too long and therefore overgraze and over-browse.

Conclusion: The answers to the questions are not as simple as just regulating the number of animals. It is particularly difficult for wildlife managers to control time and determine what the land will carry.

If we keep numbers low enough to avoid heavy die-off in dry years, then in average and good years, this leads to over-rest and a shift in succession to woody species. This, in turn, encourages the use of fire and technology to manage the crisis.

Some of the answer lies in using livestock as a tool to manage wildlife. Much can be done with animal impact.

Tool 6: Animal Impact

Animal impact includes all the things animals do besides eat—dunging, urinating, salivating, rubbing and trampling. It is critical to the management of brittle environments and useful in the non-brittle environments as well.

Two of the four insights (brittle vs. non-brittle environments and the role of herding animals) led to the recognition of animal impact as a tool.

The following are examples of the versatility of animal impact as a tool:

- In a fairly brittle, high-rainfall area over-rest allowed several years of old material to accumulate. Roots were damaged and the community had started to shift toward forbs, shrubs and trees. Fire pollutes the atmosphere, exposes the soil and invigorates woody species. Chemicals or machinery cannot guarantee that grass will establish. But, periodic high animal impact removes old material, invigorates existing plants without exposing soil, creates conditions for new plants and moves the community away from noxious weeds or woody plants. Low animal impact and/or prolonged partial rest are what cause such situations.
- In a less brittle area, a farmer wishes to maintain soil structure while he disposes of crop residues in time to plant. With very high animal impact, crop residues feed the animals and nutrients are returned to the surface as dung or urine while the uneaten portions are broken up and laid down. The animals have not been anywhere long enough to cause excessive compaction.
- Suppose you need a firebreak. A fine spray of dilute molasses or saline solution sprayed on the forage will attract and bunch a herd that will make the firebreak.
- Suppose a noxious plant has invaded. Continuous doses of very high animal impact followed by well-planned recovery periods will diminish the offending plants by moving succession beyond the stage that suits them.
- Bare eroding ground can be subjected to periodic heavy impact by giving a large herd a few bales of hay to excite and concentrate them on the area.
- Erosion gullies with steep banks provide no foothold for plants. Why pay for a bulldozer to chew up more land while consuming diesel and polluting when you can break down the sharp edges and create conditions for plant growth with high animal impact? It can also be used on the catchment to correct the poor water cycle that caused the damage in the first place.
- Low densities of cattle will not touch impenetrable brush. But, the smell of molasses blocks thrown into the thickets will attract a herd. They will penetrate the thicket, break down the brush and open it up to sunlight so the grass can grow.
- Fish management often requires steep, vegetated banks rather than steep eroding ones. Very high animal impact for very short periods can promote this.
- Stock trails threaten to wash out. Damage caused by trampling can be cured by trampling because of the vast difference between prolonged, one-way trailing and the milling of bunched animals for short periods.
- Coarse, fibrous grass dominates a bottomland that has been managed with low stocking rates and partial rest. Traditionally, fire has been used to "keep the grass palatable." But, this pollutes the atmosphere, exposes the soil and, when combined with low animal impact, leads to wider plant spacing and even more fibrous plants. High animal impact will remove the old plant tops, cover the soil and favor more lateral growth, closer plant spacing and less fibrous plants.
- Desert soils are hard capped from lack of disturbance for over 3,000 years. Nomads and their herds continue to overgraze and partially rest the land. Once a herd is concentrated and bunched briefly, the desert starts to recover.

Animal impact is the only tool that can realistically halt the advance of deserts over billions of acres of rough country.

The Role of Livestock: No other tool has sparked the level of controversy as has animal impact. It is a deeply held belief that livestock trampling damages both plants and soils. For years, academics have reject the single idea that has the most promise of stopping desertification while developing machines of extraordinary size and cost in an attempt to achieve the same end.

Since we have lost most of the large herding wildlife species and their predators, we are left only with livestock to simulate that role. There is no other tool that can both restore and sustain healthy grasslands in the brittle environment.

Unfortunately, livestock are usually seen as the enemy of the land. The recent concern over methane production by cattle is ridiculous. There were previously a great many more grazing ruminants on the Earth than there are today and all ruminants produce methane gas. Something other than cattle is responsible.

To date, most of our experience with animal impact as a tool has been confined to domestic livestock.

What Animal Impact Does: We use animal impact as a tool for the three things that it does:

1. Hoofed animals tend to compact the soil by concentrating a big weight on a small foot.
2. When animals are excited or closely bunched, trampling causes breaks and irregularities on the surface of the soil
3. Animals speed the breakdown of plant material through dung and urine and also by returning uneaten old plant material to the soil surface through trampling

Whether any of this works for good or ill depends on the management of time.

A simple example will illustrate the importance of time management.

Let's say you have a house on the hill and a donkey you use to fetch water from the stream below. You make one trip per day. At the end of a year, you have 365 donkey days of trampling. Now suppose that you took 365 donkeys down the hill and fetched a year's supply of water in one morning. You would still have 365 donkey days of trampling. But, the trail would have 364 days to recover and become greener and healthier than before. In other words, it is *time*, and not animal numbers that is the critical factor.

Stock Density and Herd Effect: Herding animals behave in a variety of ways that produce different effects. *Stocking rate* refers to the number of animals continuously supported by a unit of land. *Stock density* is the concentration of animals on any sub-unit of that land at a given time. Neither describes how they are behaving—spread or bunched. *Herd effect* is the result of a herd that is trampling because they are bunched or excited.

Trampling pushes down dead plant material and disturbs the surface of the soil. It is a result of behavior and is different from the effect animals have when they are walking calmly. Herd animals bunch when they are threatened by predators, being driven or jostling each other when given hay or supplements. It also occurs when they are in ultra high densities (1,000 to 2,000 per acre or more) and moved every few hours. Even when they become accustomed to it, their impact remains high because of their density.

Animals will normally avoid coarse plants and place their hooves carefully. When herd effect occurs, they trample these plants down, break the surface cap on the soil, and compact it enough to provide seed to soil contact.

Land is partially rested when animals are present but do not produce herd effect.

Conventional range managers advocate protecting the soil cap and its algal crust because it does prevent erosion to a small degree. Breaking the crust will increase erosion in the short term. It takes long-term observation to see that it establishes plants that protect the soil much better than an algal crust.

Over-trampling: Trampling carried out too long damages soil and plants. Temporary over-trampling (cattle concentrating in a fence corner during a heavy rain storm) seldom has a lasting effect. The area will look much better after a season. Continuous trampling around gates, water and feed bins does not allow for recovery. Even careful time management may not eliminate it all but the area right at the water can be treated as a sacrificial area.

On Croplands: Animal impact used periodically on croplands can be seen as a form of biological tillage. It also saves the cost of applying manure and other fertilizers.

The American Dust Bowl was a result of the failure to maintain the organic components of the soil and ground cover in the absence of herd animals. The development of perennial grain crops could help us get away from the destructive effects of annual shallow-rooted grain monocultures. These will require the removal and recycling of the old material that does not decay in the brittle environments. High concentrations of livestock for very short periods can speed the cycling and decay of these crop residues.

Forests: Forests that have long been subjected to numerous small fires lit by humans have changed from fire sensitive to more fire-dependent species in response to the burning. Management tends to alternate between controlled burning and doing nothing. Neither will maintain the few remaining fire-sensitive forests. Animals can clear the understory without damaging the soil surface or reducing soil organic matter content.

The African teak forests evolved with a large animal component. The tools currently being applied (over-grazing and partial rest in conjunction with fire) are destructive.

Animal impact can replace fire. But, it will lead to a decline in fire-dependent species. Therefore we must be clearly aware that we are making a rational management choice.

Conclusion: One of the most beneficial uses of animal impact in any environment is for the restoration of water catchments. There is no technology that can replace animal impact in brittle environments.

Very Brittle Environments

Community Dynamics: Animal impact promotes the advancement of biological communities on bare ground and maintains grasslands preventing them from sifting to woody species. Low animal impact (partial rest) has the same effect as total rest. It will produce bare ground and increase plant spacing and move the community toward forbs and woody vegetation.

Water and Mineral Cycles: High animal impact improves while low animal impact reduces their effectiveness. When applied along with overgrazing, the adverse effects are compounded.

Energy Flow: High animal impact builds community complexity, improves water and mineral cycles and, consequently, energy flow improves. Low impact reduces energy flow which is compounded by over grazing. This is the most pervasive situation in the world.

Non-brittle Environments

Community Dynamics: High animal impact maintains grass root vigor and discourages the shift to woody communities but may not entirely halt the movement back to forest. So, if grassland is intended, the application of technology may have to be considered. Non-brittle grasslands that cannot advance to forests because of shallow soil, elevation or some other factor increase in complexity with animal impact. Low animal impact (partial rest) has little effect on these grasslands. Woodlands will develop if the climate allows. When the community cannot develop due to some limiting factor, low animal impact does not cause deterioration and will not produce bare ground in a non-brittle environment.

Water and Mineral Cycles: Animal impact improves both. However, it tends to maintain grasslands where these cycles are not as effective as they would be if the community advanced to forest. Low animal impact has little effect.

Energy Flow: Animal impact increases energy flow but again, when used to maintain grassland, energy flow will not reach the full potential of the land. Low animal impact has little effect.

Tool 7: Living Organisms

By using the term *living organisms* (instead of *plants* and *animals*) we are forced to consider bacilli and viruses.

The two tools of grazing and animal impact involve using living creatures in management. But a separate heading of *living organisms* encourages us to consider biological solutions in lieu of technological ones. It also encourages our treatment of the complex of life as a whole rather than a list of pesky or beneficial creatures that we can kill or husband at will. Failure to think along these lines is responsible for much of the environmental damage man has done. People in less developed cultures live closer to the land and use simpler technologies and have instinctively used living organisms as tools for a long time.

Living Organisms and Community Dynamics: The living organisms tool involves all life. Therefore, it plays a role no matter the type of business being managed. The ecosystem process of community dynamics and the living organisms tool represent two aspects of the same thing. The dynamics of any biological community is manifested in living organisms.

The earliest cave dwellers tool boxes were nearly empty. They had no fire, no livestock and no technology. They knew intuitively that community dynamics was in absolute control and defined what they could do. When they harnessed fire and made their first spears, they initiated many of today's deserts. With increasingly sophisticated technology, we continue to think we can escape the influence of community dynamics. We cannot.

Practical Applications: When the Chinese noted that the Gobi Desert was advancing by over 600 square miles each year, they begin planting trees—i.e. an overt implementation of living organisms as tools. There are many other examples. For instance:

Japanese scientist, Masanobu Fukuoka used principles of biological succession to produce high yields of small grain crops without synthetic fertilizers, compost, pesticides, soil disturbance or weeding. His understanding of community dynamics allowed him to apply a number of plants, insects, birds, small animals and microorganisms.

A Mexican rancher built a concrete ramp up the outside and down into a water tank. One night he discovered a massive mating of toads in the tank that, by dawn, had dispersed in pursuit of bugs and flies. The ramp enabled the complexity and stability of the whole area to improve.

Then there was the case of a Namibian rancher who ran water pipes along his fences to frustrate the local plastic eating porcupines. But, baboons would still bite through the plastic and cause leaks. Half drums placed below the leaks provided water points for thousands of birds, insects, and small mammals.

The hope of using technology to protect our crops and reap their bounty independent of community dynamics is, and always will be, a false hope. However, the arrogant assumption that technology can replace natural law is so deeply embedded in our culture that it distorts scientific reasoning. The use of biological controls (in lieu of chemicals) is a more positive marriage of science and the knowledge of how communities function.

Genetic Engineering: Breakthroughs in genetic engineering have opened the door to tremendous opportunities but also equal dangers and temptations. We must avoid embarking on a new Green Revolution as faulty as the last. For example, attempts to escape natural law by creating crop plants that survive herbicides used to kill all other plants represent the wrong kind of thinking. We have been so enamored with technology that we have ignored the web of relationships that define a biological community.

Tool 8: Technology

The belief that technology is the hallmark of modern humans and holds the key to the future has been held for thousands of years. The 20th Century saw a constant stream of developments that fostered this belief.

Current use of technology considers only the problem at hand without thought given to the larger implications. Many of the products we use daily affect the environment in ways never anticipated. Most of these hazardous inventions have existed less than a hundred years. However, we cannot revert to subsistence agriculture nor can we abandon our cities. The future will demand wisdom.

The answer is *not* collectivism (which will lead to the greatest tragedy of the commons of all time) but the protection of private property rights and a return to truly free markets.

Technology and the Quick Fix: Quick fixes can prove costly in the long term. This is especially apparent when technology is used to increase the productivity of deteriorating land or to drastically modify an environment.

Agriculture's Addiction to Technology: There are two often overlooked attributes of nature. First, the ecosystem is a living thing that reproduces itself according to its own principles. Second, the life that we artificially suppress may contribute to our own survival. Ignoring these attributes triggers the same mechanisms of dependency as drugs and alcohol and, like any debilitating dependency, the habit quickly becomes expensive. For example:

In 1973 it took 53.3 kg of corn to buy 100 kg of nitrogen fertilizer but within a decade it took 527 kg. Corn grower's cash flow could not stand a cold turkey withdrawal because the dying soil would not grow enough to pay last year's debt. They became desperate to raise money for the fix.

Developing a Collective Conscience: If your house is on fire, no one expects you to just watch it burn. But, you should not throw kerosene on it either. But this happens sometimes—dams and wells have helped scatter livestock thinly on lands all over the world.

We cannot hope to feed the world's future population without technology. But, we can surely avoid applying it pathologically.

Alan Savory writes: "At present we have no …collective sense of conscience and responsibility, either to our fellow humans or to other life, and our governments reflect this only because our governments reflect us…"

That is *absolutely untrue*. Governments do not reflect "us." "We" are not the government. No matter the "ism" you put behind it, all governments are nothing more than a group of men who hold a regional monopoly on the use of force and violence (a monopoly that they probably acquired through the use of force and violence). *Governments are the problem.* Not only are they the single largest polluter, they are the primary source of funds for technological research and education – funds that were taken from someone else at the point of a gun.

Private property rights and a totally free market (completely unhampered by government intervention by any means) are the only solution to these problems. Furthermore, privately funded education is the only means by which the current paradigm will ever change enough to make this solution work.

Lesson 5: Management Guidelines

Guideline 1: Controlling Time

To restore and maintain a biological environment, you need to minimize overgrazing of every plant you can. To do this, you base the time plants are exposed to animals on the most severely grazed plants.

Adverse consequences of trampling are also a function of the time the land is exposed to animals and not absolute animal numbers. Maximum impact over minimum time followed by a sufficient recovery period makes trampling very effective in maintaining healthy soils.

Monitor the Perennial Gasses: Perennial grass species are the most vital to the stability of the whole community. Especially in the low rainfall areas, they contribute more to the health of the whole community than any other factor. So in these environments, grazing should be timed to the needs of the perennial grasses.

Monitor Plant Growth Rates: A plant is overgrazed when it is severely bitten and severely bitten again while using energy from its other parts to reestablish leaf. This can happen in the grazing period when animals are left too long in one place. It can also happen following the recovery period when animals are returned to an area too soon. A plant can be safely re-grazed when all its roots are re-established.

How long is too long and how short is too short depends on two things: 1) the proportion of leaf removed and 2) growth rate. The faster the growth rate, the shorter the grazing period needs to be. When growth is fast, the shorter the recovery time needed. With runner type grasses this can be as short as 12 to 15 days while with bunch grasses it may take 25 to 30 days. When growth is slow recovery times for runner type grasses can stretch to 30-50 days and bunch grasses from 60 to 120.

Grazing and recovery periods are linked. Grazing time in one area will change recovery time in other areas and vice versa. Each area that is grazed for fewer days than planned reduces recovery time in all areas. Conversely, each day livestock are held longer on an area adds a recovery day to all other areas.

Base Grazing Periods on a Preselected Recovery Period: Land that is managed as a unit is called a cell. Subdivisions of that cell are called paddocks.

Perennial grass plants in most areas of the Southwest need a recovery period of 60 days during slow growth. So, if a cell contains 9 equal paddocks, a 60 day recovery period would require a 7.5 day grazing period. (60 day recovery divided by the 8 paddocks that are not being grazed.) Note that, as the number of paddocks increases, the length of the grazing period decreases.

Advantages of Many Paddocks

Increasing paddock numbers decreases the time spent in each paddock and therefore increases your ability to minimize overgrazing. Furthermore, as paddock size decreases, stock density increases which means improved distribution of dung, urine and trampling.

More Even Grazing: As paddock size decreases, the proportion of plants grazed increases. This does not mean that animals are any less able to graze selectively. The time they spend in a paddock also decreases when paddock size decreases. Therefore, the same volume of forage is taken from the cell as a whole as before the decrease. The only things that can change the total volume of forage taken from a cell are the number of animals or the time they spend in the cell as a whole.

As the animals select for a balanced diet, they feed over a larger proportion of the plants available. This keeps a higher proportion of the leaf on more of the plants fresh and young.

Increased Energy Flow: The amount of green leaf removed greatly influences the rate at which plants can re-grow after being grazed. For example, assume that plant A is 90% defoliated and plant B is 40% defoliated. Over the next couple of weeks, B will produce a greater volume of new leaf while A struggles to replace lost energy reserves in its roots before it can produce new leaf. It will take awhile but, at some point A will catch up. So, the higher the proportion of less severely grazed plants to severely grazed plants, the shorter this "catch up" period will be and the more total forage will be produced in a given recovery period.

Thus, in practice, the more paddocks per cell, the better the distribution of grazing, the fewer severely grazed plants and the greater the proportion of plants able to recover quickly—all of which increases energy flow.

Improved Animal Nutrition: Animals that move frequently onto fresh, un-fouled ground receive a better plane of nutrition and are in less danger of parasite infection.

The first day in a paddock, animals select what is readily available. They do the same the following day but do not find it quite so easy and so on as they experience a lower quality diet with each passing day. If they stay in a paddock until the forage is depleted, the drop in nutrition results in poor performance.

Time management becomes ever more critical as paddock size decreases. At some point, even a 24 hour mistake can mean extreme forage depletion and commensurate drop in animal performance.

In short, more paddocks decrease the risk of weather fluctuations but increase the penalty for poor management.

Time and Over-browsing: Timing grazing to the needs of perennial grasses brings up the question of bias against woody plants.

Academics did not (and some of them still do not) believe that severely defoliated desert brush could not recover in short recovery periods. Let's see.

With sixteen paddocks and a 60 day recovery period the grazing period is 4 days. This reduces stress on animals, cuts trampling time and ensures a good chance of achieving a grassland. Further, it does not expose bushes to the kind of heavy and prolonged browsing that requires long recovery periods. In addition, the increase in perennial grasses and other plants lowers the pressure on shrubs.

Indeed woody plants can thrive when the length of the recovery period is designed for perennial grasses.

Grazing in the Dormant Season: Perennial grasses are not susceptible to overgrazing when they are dormant. However, when herds linger in the same area they will avoid dung and urine fouled ground, their hooves continue to trample and incidences of parasites and infection will increase.

Furthermore, dormant periods are critical for wildlife. Arbitrary rotation of livestock can devastate wildlife due to the decreasing plane of nutrition. This will also cause livestock condition to suffer and/or result in very high supplemental feed cost.

Limit the Number of Selections: Regardless of season, when animals enter a fresh paddock they balance their diets as best they can. But, during a second rotation in the dormant season, plants will not have regrown any new leaf. However, the impact on animal nutrition will not be the same as one prolonged stay. The recovery period will allow fouling to wear off and moving livestock on to fresh ground stimulates them.

Unquestionably, the forage remaining the second, third and fourth rotations will be less nutritious and more fibrous. Therefore, aim to limit the number of times animals have to select from the same paddock and move as frequently as possible. The more paddocks you have, the greater your ability to accomplish both. For example, with a 200 day dormant period and 100 paddocks and a 2 day grazing period, would only require one selection. With only 10 paddocks, you would have to select over the same forage several times. Stock would require supplementation toward the end of each grazing period in order to maintain rumenal microbe populations.

If you are concerned about the inhibiting influence the high amount of fencing required would have on wildlife, herding or strip-grazing can be substituted.

Plan a Drought Reserve: Planning for drought by withholding certain areas from grazing during the growing months will decrease the productivity of both the forage and the animals. It is much better to plan for drought by reserving *time*—i.e. we measure drought in days.

Say, for example, your normal dormant season is 200 days. Add 30 to 90 to that and then plan the frequency of moves through the dormant season on that higher figure. Calculate the number of animal days available in every paddock and ration them out carefully.

Time and the Management of Wild Grazers and Browsers: Any number of animal species can provide trampling in either a beneficial or a detrimental way if they stay too long or come back too soon.

With livestock we can distinguish between time and numbers because we can control both. But, wild animals are not subject to that kind of control. In their case, numbers can influence time. The social behavior of large herds shows little resemblance to that of small groups.

Where predation, accident and disease control animal numbers, the size of the animals' home range regulates the frequency with which they return to feeding areas. Concentrated fouling insures sort periods of grazing. If the removal of predators or other causes of death permit numbers to rise, home ranges become smaller as more herds occupy the same area and return to past feeding grounds sooner. This starts the breakdown of the ecosystem including the loss of non-herding species. It is what causes the ecosystem to break down so badly after game reserves are formed.

In the 1950s Vesey-Fitzgerald observed what he called "grazing succession" in Tanzania which provides a clue for the use of livestock to induce movement in other species. The game came to his area in a definite pattern. Elephants that can digest very coarse, tall and fibrous grass came first. Next came grazers that could handle the coarse forage that had been opened up by the elephants—zebra and buffalo. Finally the smaller species came in.

More recently, a ranch in Zimbabwe ran 60,000 head of cattle and a game ranch. Paddock fencing was designed to constrain cattle but allowed game to move freely. Game routinely turned up in paddocks two moves behind the large herd of cattle.

Of course, the arrival of game after the stock increased the grazing period and shortened the recovery period. However, the high animal impact overcame the slight periodic overgrazing and the land improved dramatically.

Fortunately, control of time does not stand alone and perfection is not necessary. Simply do your best to minimize overgrazing by carefully planning time. Although some will surely occur, high animal impact can overcome that.

Guideline 2: Stock Density and Herd Effect

Stock Density is the number of animals on a given area at a given time. For example: If 100 animals are in a 100 acre paddock, stock density would be 1 to 1. Move them to a 200 acre paddock and stock density would be 1 to 2.

Herd Effect can't be quantified. It is the effect on soils and plants that a large number of animals have if they are bunched so closely that their behavior changes—the larger the herd, the greater the effect.

Low stock density (and the partial rest that goes with it) characterizes conventional range management and is to blame for the serious degradation of the world's rangelands. Trailing and a high degree of "patchiness" (patch grazing) is its marker. The term *low-density* grazing describes the process and suggests a solution which is developing many paddocks or strip grazing within paddocks.

When stock density gets higher, time on the land gets shorter, dung, urine and trampling are more evenly distributed and animals move more frequently to fresh pasture. Be that as it may, animal performance is good reason for increasing stock density for its own sake.

Stock Density and Animal Performance: The early days of planned grazing were plagued by continuous poor animal performance. The problem stemmed from the conventional wisdom that animals select their diets by species that are more palatable and reject other less palatable species. But that did not (and does not) explain why it is easy to observe a plant (of the "less palatable" variety) grazed completely to the ground standing right by the side of a plant of the same species that is untouched—same species, same soil, same weather, same exposure to animals.

Animals do *not* select by species. They don't even know the Latin names. They select the freshest and leafiest forage on *any species*. Cattle carefully select their diet from what is in front of them. They will eat the fresh tender leaves of undesirable species A and leave the old oxidizing leaves of species B alone.

In smaller paddocks with animals at a higher density the plant community has more leaf and less fiber. Increasing stock density by subdividing large paddocks into smaller ones and combining several herds into one will generally improve animal performance. Once stock density is increased, rainfall, soil type and the manager's ability to plan and monitor are responsible for any variation in animal performance.

Rainfall and Soil Type: The soils in low rainfall areas are generally high in mineral content and more alkaline. The forage they produce is generally shorter but has less fiber and cures better. The higher mineral content helps keep rumenal microbe populations high and thus maximizes digestive efficiency.

By contrast, soils in higher rainfall areas are more acidic and produce taller, tougher forage with much higher fiber content. Without extensive supplementation rumenal microbes decrease and livestock perform poorly.

In the early days of planned grazing, we found that we could increase stock density immediately in the low rainfall areas and experience little or no drop in animal performance. However, in the high rainfall areas the old forage (and inevitable initial drop in performance) had to be burned or mowed or we had to supplement heavily or accept the stock performance decline as a legacy of the past. We tried to mitigate the latter by grazing and trampling down the old stale forage at a time when the animals could drop in condition without the rancher suffering a financial loss.

Furthermore, in the low rainfall areas we found that we could get by with starting at a lower density and increasing it. The typical result was cells being patch grazed at least a part of the year. During rapid growth (with recovery periods based on how long it would take a severely overgrazed plant to recover) many plants would go un-grazed. Then, when growth slowed and movements slowed down the animals would start to graze the un-grazed plants because they were still fairly nutritious. Then, when growth stopped and we began to ration out the remaining forage, the un-grazed patches would generally be cleaned up with animals dropping only slightly in condition.

Areas of high rainfall were more perplexing. As expected, performance suffered when paddocks were few and stock density low even though stocking rate had been doubled. During slow growth (when grazing periods were lengthened) animal performance would drop because much of the available forage had not been grazed previously and had lost its nutritional value. The patchiness was even greater because of the volume of forage produced by the high rainfall. Then by the end of the growing season, a large proportion was rank and of little use in sustaining animals through the dormant season without heavy supplementation.

What was learned: In order to attain good animal performance in high rainfall areas, try to get to 100 paddocks (or strip graze within paddocks) as soon as possible. Your goal should be a density high enough to ensure animals will graze or trample a high proportion of the plants during the growing season to keep them fresh and nutritious in fast or slow growth and, thus, more capable of sustaining animals through the dormant season.

Again, in high rainfall areas, a primary indicator that you can monitor during the growing season is seed heads. Plants have two stages of growth—vegetative and reproductive. The plant's goal in life is to set seed and reproduce itself. The grazer's goal is to prevent it from doing that—i.e. to keep the plant in a vegetative (and hence more palatable and nutritious) state as long as possible. So (assuming recovery periods are properly calculated) when it comes time to make a scheduled livestock move, if you have a high proportion of developed (or developing) seed heads in the paddock you are leaving, you do not have enough animals.

Grazing Planning: The poorest results always occur among managers who fail to monitor plant growth rates and properly adjust grazing and rest periods. Problems most commonly occur when growth rates slow down but grazing and recovery periods are not adjusted accordingly.

Rapid moves during slow growth provide a short-term benefit to the animals. However, the associated shorter recovery times mean the animals return before the plants has time to recover. Thus, the plants that were grazed severely in the previous grazing period are overgrazed and those that weren't grazed at all are left to grow even staler. Ultimately (in the longer-term) this leads to animals being forced to graze the stale forage because the overgrazed plants are unable to produce enough to feed them. In low rainfall areas, performance may drop too little to gain immediate attention. But, in high rainfall areas, stock stress appears almost immediately.

Both overgrazing and over-resting are minimized by moving animals according to growth rates. Slowing moves down with slower growth rates lowers individual animal performance. On the other hand, continuing fast moves benefits the animal in the short-term but damages both the land and the animals in the long-run.

Ultimate conclusion: If you wish to manage and maintain a grassland, a high stock density is required at both ends of the brittleness scale.

Herd Effect: Herd effect means high animal impact. It is produced by a change in animal behavior to bunching, milling or excitement. Neither wildlife nor livestock produce much herd effect without an outside stimulus. In truly wild herds the bunching-milling of large numbers of animals was their most effective protection against predators.

Where predators caused bunching and the forming of large herds, the concentrated dung and urine induced movement which, in turn, regulated the time plants were exposed to the animals and thus prevented overgrazing.

When animals spread out to feed, they avoid stepping on coarse plants and their hooves do not break the soil surface or trample old plant material. Without pack-hunting predators, most herding animals spread out into smaller herds.

Most anyone would understand the impact of withholding rain. But, the damage to water cycles caused by eliminating herd effect (and replacing it with fire) has done exactly the same thing.

Stock density is a function of paddock size and animal numbers. However, herd effect results from animal numbers and behavior, regardless of paddock size. Animal impact is applied by using stock density and herd effect.

Numerous researchers who have no understanding of herd effect have, in effect, proved that low stock density does not do what we claim high herd effect does. To elaborate: Two steers in a one-acre paddock will not have the same effect as two hundred steers bunched for a time within a 100 acre paddock, although stock density is the same. A great deal of money has been wasted over the years studying low animal impact over prolonged time instead of high animal impact over short periods of time.

Practical Demonstrations of Herd Effect: We encourage both researchers and stockmen who fear trampling (and hence herd effect) to conduct simple experiments of their own. For example, enclose 400 cattle on 5 acres for only a few hours—long enough that nearly every plant will be grazed or trampled down. More commonly the enclosures are smaller (one acre or less) in size and the number of cattle range from 300 to 1,000 for anywhere from an hour to a few hours. The intention is to achieve maxim density for a minimum time. Keep in mind that a wild herd would have remained on a particular piece of ground for only a few minutes out of many months or even years.

Inducing Herd Effect Routinely: Fencing and grazing systems that spread livestock thinly over the land are unnatural and have exacerbated the problem. Predator-induced behavior must be simulated. This can be done through training, attracting animals to an edible reward, herding and or temporary electric fencing to push stock density so high that animal behavior changes.

The Type of Livestock Matters: Almost any type of livestock (even sheep and goats) can produce adequate impact on sandy soil. However, on hard surfaced tropical clays only cattle or horses will do the job. Also, in very tall old grass clumps and for opening up very dense brush, large animals like cattle are necessary.

Herd Size Matters: There is no evidence of any drop in performance of a very large herd with good handling facilities, calm handling and well-planned grazing. The larger the herd the better is especially true in brittle environments. Herds of 2,000 to 5,000 head followed by longer recovery are much better than small ones of 200 to 500 followed by shorter rest periods.

Using Attractants to Induce Herd Effect: Of the two methods for inducing herd effect, using attractants is the simplest. Range cubes, a bale of hay or a bit of granular salt will work. Supplement blocks or molasses/urea liquid do not work as well because they do not excite the animals which will come a few at a time and linger too long in the vicinity.

You can easily train animals to come to you anywhere you have scattered an attractant if you blow a whistle each time you do.

This technique's one major drawback is that very little ground is impacted at a time. Two thousand head will affect an area of only about 50 yards across each time they are attracted. In addition, it is very seldom practical to induce a herd more than once a day.

Using Ultra-high Densities to Induce Herd Effect: Using a combination of herding and portable electric fencing to strip-graze very small areas of land can achieve stock densities of 1,000 to 2,000 animals to the acre with recovery times as long as 200 days or more.

The behavior of the animals changes at such high densities. Although one might expect otherwise, animal performance does not suffer unduly even in a breeding herd that includes small calves.

Herd effect is greatest when animals are first introduced to grazing at such high densities and decreases as they grow accustomed to it. However, it still remains effective and is a practical (and profitable) way to increase herd effect over millions of acres of rangeland.

Guideline 3: Burning

Advocates of burning fail to understand that the benefit comes from disturbance in any form, not just burning. Animal impact can be used to achieve the same end without the adverse effects of fire.

Common justifications for burning:

- To invigorate senescent grasses (if animals cannot be used for some reason) or to maintain fire-dependent species
- To invigorate and thicken brush for wildlife cover
- To expose the soil in patches to create a mosaic that can support a diversity of species
- To reduce fire-sensitive woody plants
- To provide intense disturbance on an area with many dead plants that are hindering growth

Before You Burn: If you burn to rectify past management, be sure you also rectify the cause. If you don't, you will end up fighting the past effects of fire with fire. Although fire may freshen individual plants, it exposes the soil between the plants. Over time, this will result in fewer and larger plants that become coarser and more fibrous.

Forbs that spring from bare exposed soil are the beginning of an advance in biological succession. Burning them sets the process back to square one.

Weak Link (Social): The currently prevailing belief is that fire is natural. Therefore it is not likely that your burning will offend anyone.

Weak Link (Biological): To reduce the population of fire-sensitive species, you need to be aware of the weakest point in the species' life cycle. You would be likely to kill adults and reduce seedling establishment of plant species that establish best in long-rested perennial grass plants.

Weak Link (Financial): Burning to reduce fire sensitive woody species is tempting in years of low forage production. But, low forage production indicates that energy conversion is the weak link in which case it would be a mistake to burn. Burning is best left to years when forage is abundant and energy conversion is not the weak link.

Marginal Reaction: We erroneously view fire as being cheap. But, when lost forage and grazing and the reduced effectiveness of rainfall are considered, the true cost is high. Plus when atmospheric pollution is factored in, the cost is even higher.

Gross Profit Analysis, Energy/Money Source and Use: A gross profit analysis would not normally apply nor is the energy/money test likely to affect most burning decisions.

Sustainability: Look specifically at the future resource base described in your strategic goal and consider the following: Soil exposure, litter loss, effects on mineral and water cycles, the microenvironment at the soil surface and potential effects on community dynamics. Think about the entire future community and its age structure.

Burning to eradicate brush almost always fails the sustainability test because fire actually invigorates many woody species. It also exposes the soil and produces long-term damage.

Also, you should consider how a burn might affect your neighbors if it got out of control.

Society and Culture: When you burn you release carbon and other pollutants into the atmosphere. You contribute these pollutants each time you burn when there is an alternative or when you burn a large area when a smaller one would do.

Planning Considerations: How you plan to burn is just as important as why.

Prior to Burning: You can use animal impact to create fire breaks by spraying a thin molasses-water or salt-water solution in strips.

Types of Burns: Burns may either be hot or cool depending on fuel, moisture content and humidity.

In tropical areas with wet-dry seasons, the best time for a hot burn is toward the end of the dry season.

In temperate areas, the best time depends on several factors. Cool burns are done when forage is still partially green and difficult to ignite. When these are done at the beginning of the dry season, the soil will remain exposed longer than with a hot burn done later in the season. Hot fires burn uniformly while cool burns are usually patchy.

If the situation calls for a hot burn for uniform burning and shorter soil exposure, you must ensure sufficient fuel by not grazing it down. If you need a hot burn to kill a fire-sensitive plant, then burn when that plant is in its most vulnerable stage.

On the other hand, a cool burn requires less fuel so you may be able to take some grazing out before the burn.

Tools to Associate with Burning: Fire becomes fire followed by rest by default. That is the standard practice around the world. Both produce bare ground. Animal impact offsets the rest and is the tool of choice for use after fire. Wild grazers concentrate on burned areas, often moving onto these areas before the ground has cooled off. It can be induced with livestock by using an attractant and/or a single strand temporary electric fence to increase stock density.

Burning to Enhance Wildlife Habitat or Reinvigorate Biological Communities: Since many plants and animals are fire dependent, it may be a mistake to suppress fire altogether. Excessive old plant material will need a hot burn if, for some reason, it is not possible to use animal impact. However, the patchiness produced by cool burns increases edge effect and hence the most varied wildlife habitat. But on the other hand, cool burning repeatedly will eventually lead to a great increase in the fire tolerant woody species and the loss of that patchiness.

Monitoring: There are a couple of things that you should always keep in mind: First, frequent use of fire is detrimental. Second, the drier the climate the more dramatic the effects of fire and the less frequent it should be used.

Furthermore, each time a decision is made, assume that it is wrong and decide on what you should monitor that would give you the earliest indication that you are moving away from your goal. Since the surface of the soil is the key to managing all four ecosystem processes, start there. Another indicator is the type of species of plants that appear after the burn.

We must manage for the health of the ecosystem as a whole, not particular species. Therefore, monitor the factors that affect all four ecosystem processes and provide the earliest indication of change —for example, liter cover, soil exposure and plant spacing. Also, changes in the age structure of plant species is probably the best indicator of what is happening to community dynamics.

Conclusion: Monitoring the soil surface helps determine how frequent you should burn. Fire every 20 to 50 years can be beneficial. Fire every two to five years will lead to catastrophe.

Guideline 4: Population Management

The management of community dynamics applies when we want to encourage or discourage species succession.

Any population that does not reproduce will disappear without any harvesting at all.

Two kinds of knowledge are needed to manage living populations: First, asses its health and stability. Then, pinpoint the cause of its condition.

Techniques that reduce everything to the same age and on the same schedule do violence to the natural dynamics of communities and inevitably lead to instability and ultimate failure.

Self-Regulating and Non-Self-Regulating Populations:

With respect to population dynamics, there are two types of animals--those that regulate their own numbers and those that do not. Self-regulating populations manage to limit their numbers even though they have very high breeding rates and a potential for rapid expansion. If we protect them they do not increase. If we try to shot them out, they breed as fast as they are shot. They usually present little difficulty in management.

On the other hand, non-self-regulating populations may appear to be self-regulating as long as they exist in complex biological communities that remain relatively stable. Herding, flocking and other gregarious species fall into this category. Their populations can explode into problem numbers if the integrity of the community as a whole is damaged. Once predators are removed, populations can face a heavy die-off or become severe pests.

Age Structure and Population Health: The S shaped curve describes the growth of most all populations. At the beginning the population increases gradually. Then it accelerates geometrically and finally falls of as numbers approach the community's biological capacity to sustain them. The importance of the individual changes as the population grows. At the outset each seedling or breeding pair has a large impact. But, the impact of the individual becomes less as the geometric progression advances.

Environmental Resistance: This is the term used for the limiting pressures exerted by the whole biological community on a population. When the built in checks and balances (like predators) are removed, the population may explode. This, in turn, pressures other populations and destabilizes the whole community.

At each point on the S curve, the population has a characteristic age structure.

Age Structure versus Numbers: Age structure reflects where on the S curve a population lies. It provides more useful information for management than does the numbers of individuals. Accurate field counts of wild populations (even of immobile plants) are nearly impossible with current techniques and fail to meet acceptable scientific standards.

When major populations show no significant survival of young, bad trends already afflict all four ecosystem processes. Numerical count will not reveal this but a sample of age structure does.

Long rested, relic of pristine range sites are used by government agencies as the standard against which management of other sites is measured. These sites supposedly reflect the potential and they do, in fact, contain the "key" (desirable) species but every plant is dead or senile after years of total rest. Not a single young plant is to be found. Using such sites as the standard renders most government statistics for range management success suspect at best.

The Limitations of Game Counts: The management of game provides many examples of the limitations of counting. Numbers estimated from census seldom come close to the actual numbers present. Management relies on counting done by one or two people over a few days and thousands of acres of heavily vegetated land.

The mystique of aerial counting has been proven hollow. Large masses of animals can render themselves practically invisible from the air. There is good reason to mistrust areal counts more than any other technique.

There are a great many other questions that deserve more attention than game count numbers in the management of game. Besides age structure, there is the sex ratio in adults; feed, cover, and water requirements; home ranges; utilization of feed plants; the age structure of those feed plants; and so forth.

Bottlenecks: All creatures must be able to satisfy basic needs for food, cover and water throughout the year in order to survive. If any one of these cannot be met for even a short period, the population will be limited. This is a bottleneck or limiting factor.

When trying to increase the numbers of a species, first make sure you address the weakest link in the species life cycle. If numbers still don't increase, look for a bottleneck.

Water during short but critical periods is often the bottleneck. If water is lacking for several days during a critical time of year, the entire population may die.

Having water present is sometimes not enough. It has to actually be available to the critters. Steep-sided troughs on sites of maximum disturbance and no cover may actually kill them. If water levels are low, creatures may be able to clear the sides but drown when they can't get back out.

The bottleneck principle applies equally to plants. For example, frost limits community development as can the absence of a particular trace mineral.

Dealing with Predators that Become a Problem: Typical government policy toward predators reflects an ingrained lack of understanding of their role. And indeed, more research is needed. Meantime, government agencies go to great lengths to kill predators and make little effort to live with them or protect livestock by other simple means.

Individual predators learn to be increasingly cunning as attempts to kill them fail. Furthermore, no matter how many you kill that have not acquired the livestock killing habit, the killer remains and will educate the others. Animals learn destructive habits from each other. The particular individual must be dealt with, not the whole population.

Conclusion: The fundamental importance of the whole community can be easily overlooked when we focus on rare, endangered, or preferred species.

Lesson 6: Infrastructure Planning
Designing the Ideal Layout of Your Infrastructure

Introduction to Ranch Planning: For instructional purposes only, we separate planning methodology into three activities: Infrastructure Development Planning, Grazing Planning and Financial Planning. However, in practice, the three are treated as a seamlessly integrated whole. The student-practitioner should make sure s/he has a sound grasp of the basics of all three before attempting to create and implement a ranch plan.

Introduction to Financial Planning: A common tendency is to allow costs to rise to anticipated income. To avoid this, plan profit before planning expenses then, when planning expenses, give priority to those that will generate the most income over those that merely keep the business running.

Introduction to Grazing and Infrastructure Planning: If the role of time is properly understood, one realizes that animals have to move continuously and, to do this requires fences, water points and handling facilities. Thus, infrastructure planning and grazing planning go hand in hand.

Large tracts of land require substantial investment in infrastructure which requires planning. However, smaller tracts (for example, around the home site) seldom require a long-term plan—the infrastructure needed can usually be built within a few years and at a low cost.

Financial and grazing planning are both tactical in scope—monitoring the plan is essential and will usually lead to modifications within the year.

In contrast to conventional planning (which focuses on short and/or long-term objectives) Strategic planning focuses on achieving the strategic goal.

The practical aspects of infrastructure planning: In practice, you cannot assess the possibilities for your land plan without a background of financial and grazing planning.

A land plan is not a starting point. It is an end point toward which you build from experience and success.

Land plan implementation proceeds like biological succession—one stage makes the next one possible.

Your plan will map all future developments on the land and will affect future generations. You want a plan that goes beyond maximizing yield from one particular kind of animal.

If you follow the steps, your final plan will develop almost as a by-product of the process.

1. Gathering important information.
2. Preparing maps and overlays.
3. Deciding on herd and cell sizes.
4. Preparing maps for planning.
5. Holding the planning session.
6. Designing the ideal plan.
7. Implementing the plan.

Step 1: Gathering Important Information

A plan is no better than the information that goes into making it. So, spare no pains in gathering every scrap of background material that will help you plan wisely. Seek ideas from everyone who has an interest. People who participate become supporters.

Map Your Future Landscape: Record the broad landscape features described in your strategic goal on a transparency (clear plastic overlay).

Start a Checklist of Issues: For each item on the list, answer questions: Who? Where? What? Think in terms of categories:

- *Natural issues:* Weather; prevailing winds; areas prone to deep snow cover or flooding; geographic features; water sources and riparian areas; eroding catchment areas; differences in soil types; areas of fire threat and wildlife concerns; areas of heavy predation and areas where endangered species are present.
- *Social issues:* Present ownership boundaries; future estate plans; areas where mineral rights are leased; water rights and how they might affect your water use; future developments planned on surrounding properties; access areas for hunters or hikers; areas prone to vandalism; archaeological sites; and so on.

Prevailing winds might influence the placement of roads that can also serve as firebreaks. You would not want to build a home on a flood prone area. On public lands, putting a fence where it would inhibit movement of the public would invite vandalism.

Identify Management Factors: Review your strategic goal and be sure you address all forms of production, quality of life and aesthetic concerns. List all the management factors you can think of. Consider:

- **Livestock Production:** *Possible future stocking rates* will influence the size and number of cells and handling facilities.

    - *Herd size:* This will allow you to determine water needs in any one place. You are likely to require many more divisions of land than you have now.
    - *Access to water and other points:* Avoid repetitive movement over the same ground. Locate corridors and sacrificial areas to minimize damage.
    - *The brittleness scale:* Less brittle lands need high stock density (small paddocks) and frequent moves to ensure animal performance. More brittle environments need higher herd effect as opposed to stock density. Paddocks can be large if herd effect is induced by a very large herd using attractants or strip-grazing.
    - *Water storage and delivery.* Consider potential dam sites and locations for storage reservoirs and water tanks.
    - *Multiple species.* Running more than one class of livestock will result in better use of a wider range of forages and higher production. Though you may not run additional species for years, planning for it now saves having to modify your plan later.

- **Crop Production.** Will you run animals on certain fields at certain times? Do you irrigate? Will machinery have access? With smaller fields you get more edge effect and thus more diverse populations of insects, birds and game.

- *Timber Production.* Will you graze livestock in the forest to help with fire mitigation? What timber harvesting method(s) will you use?

Start a List of the Infrastructure Needed: Draw up a list of basic facilities and keep adding to it as you progress through the next steps.

Circulate your List and Ask for Feedback: When the list has circulated long enough to ferment, begin work on specifics.

Step 2: Preparing Maps and Overlays: A good topographical map is essential. It should not show existing fences, water points or ranch roads. These might force thinking into the old pattern. Put those features on an overlay.

- *Create a Master Map of the Property.* U.S. Geological Survey 1:24,000 scale maps are generally adequate You may have to cut and paste several of them together. Smaller properties will need a larger scale map for more detail. These can be made using a copy machine to enlarge the existing map. Aerial photos are generally too small in scale. The only constructed features that should appear on the map are those that would be illegal or genuinely impractical to change—public roads, railroad tracks and homes, etc.

- *Prepare a Map of Existing Developments:* Include water points, buildings, fences, croplands, ranch and farm roads, working facilities, etc. Then set it aside.

- *Create Overlays:* Several things can be put on one overlay as long as it doesn't become cluttered.

- *Map the future landscape:* (using the information you gathered in Step 1); Hunting areas; Winter wildlife range; Wildlife roosting, mating and nesting sites, food plants, movement routes; Deeds, leases, ad permits; Rights-of-way; Features involving multiple uses (camp sites, trails, etc.); Fire danger and prevailing wind; Crop or timber areas; Inaccessible areas; Existing facilities, fences, and roads; Water sources and water rights; Snowdrift areas; Flood areas; Estate plans (future land divisions); Mineral rights and leases; And so forth.

- *Measure Acres and Hectares:* An acre is about the size of a football field. But, football fields are hard to fit into irregular 12,000 acre cells.
    - **Acres.** Squares of overlay material (or thin paper) divided into 10 acre squares make measuring acres easy on a map.
    - One-eighth mile squared is 10 acres and 1 ¼ miles squared is 1,000 acres. You can take both of these units right off the mile scale of any map.
    - Using a 1:24,000 scale map, for example: Cut out a piece of clear plastic that is 1 ¼ miles on each side as measured by the map scale. Then divide that into 100 smaller squares each of which will represent 10 acres on the map.

Step 3: Deciding on Herd and Cell Sizes

You need an idea of cell and herd sizes to help you make decisions about land divisions. A grazing cell is the area planned on a single grazing chart and is not a particular shape or fencing layout.

As in many other aspects of management, you can't find precise answers through rigorous mathematics. But to get your thinking on the right track, consider your land as flat and uniform with water available anywhere.

The "wagon wheel" with water at the center and radiating fences is ideal where such conditions actually do exist. But, let go of the idea that paddocks must be equal and symmetric and you can shape them to fit rough country and still have a common center.

Think in terms of area per head, location of handling facilities and distance to water. A circle or square cut to scale the size of an optimum cell will enable you to see on the map how these relationships apply to your land. Your cells will likely be shaped differently. However, the thinking that goes into figuring the size of planning circles or squares (area per herd, distance to water and handling facilities) will be essential to the final design.

Making Planning Circles: Useful Formulas:

- Acres divided by 640 ac/mi^2 = square miles
- $\sqrt{\text{square miles}}$ = side of a square cell in miles
- $\sqrt{(\text{square miles}/3.14)}$ = radius of a circular cell in miles
- (Miles X 63,360 inches/mile) ÷ map scale = inches on map.

Grazing Cells and Herd Sizes: Thinking Through a Cell-Size/Herd-Size Problem:
A cell is only a piece of subdivided land that is planned as a unit. In practice, you can treat any combination of paddocks with any number of facilities as a cell. In fact, the whole property might be planned as one cell. A cell is what is planned on a grazing chart as a unit, not any particular physical layout.

Herd Size = acres in the cell ÷ carrying capacity in acres per standard animal unit (SAU).

For example, let's say that you have 12,000 acres that you are thinking of managing as a cell and this tract of land has been supporting a stocking rate of 1:20 for years or 600 SAU. (12,000 acres ÷ 20 acres/SAU = 600 SAU herd size).

12,000 acres ÷ 640 ac/mi^2 = 18.75 mi^2
$\sqrt{18.75 \text{ mi}^2}$ = 4.33 miles per side of a square cell
$\sqrt{(18.75 \div 3.14)}$ = 2.44 radius of a circular cell (miles to water)

Factors to consider:

- *Stocking rate.* Improving the land will double or even triple carrying capacity. You may have to run more animals at one time than you can easily manage just to make efficient use of forage.
- *Herd size.* Cattle, sheep and goats can thrive in herds of any size. However there are factors that offer challenges: labor, handling facilities, water, calving and weaning, etc. Thus herd size can become unwieldy at a point that must be determined only by you.
- *Number of Herds.* A single herd is ideal. But, sometime two or more herds are needed.
- *Paddocks per herd.* Ultimately, you will want many (100 or even more) because of the great flexibility it will give you and also for best performance of both land and livestock. So, when laying out paddocks, keep subsequent divisions in mind.
- *Water supply.* Frequently the limiting factor. 15 gallons/day/SAU is the commonly accepted requirement. From that, you can compute the water required. You may find that, at times, one well will not provide enough and water may have to be piped from another source. Planning now can avoid high costs later.

- *Distance to water.* This also limits cell size and can vary greatly. Livestock do not necessarily drink daily. In areas of Africa and Australia animals graze 5 miles or more from water. Plan for what is best for your situation.
- *Topography.* If you opt for a radial layout and the land slope is concave, the radials should converge toward the bottom of the slope. If convex, they should converge toward the top. This enables animals to move across flatter ground as they approach water.
- *Marginal reaction.* Development cost per acre goes down as cell size increases. But calculate carefully because: Although the same number of radial paddocks requires less fencing when arranged around several centers than when extended long distances from a single center, the cost of supplying water to each center may be substantial.
- *Land boundaries.* If circumstances warrant and cooperation seems possible, amalgamating properties for management (not ownership) may make sense.

Organize group sessions to generate a number of different plans. This brings benefits from the presence of outsiders. It costs nothing to reject a truly loony plan but never seeing a new idea can be expensive

Step 4: Prepare Maps for Planning

Get at least ten photocopies of the master map for each planning team to use for the creative planning. Make one copy for each planning team of the map showing existing developments and set those aside. The teams will create a final plan using that map after they have exhausted all possibilities using the master map.

Step 5: Holding the Planning Session

Armed with the master map, overlays, a supply of planning circle cutouts of different sizes, conduct this session when people are fresh and creative not tired at the end of a day.

Divide the Planners into Teams: Groups should not be larger than three to six people. Your goal is to get a large number of plans.

If certain people are likely to dominate others, put the bosses together. You will generate more possibilities if you can avoid the thinking being dominated by a person who "knows what cannot be done."

If you bear ultimate responsibility others are likely to defer to you without discussion. Consider not joining a team. The foreman, a local dignitary, or other leader whose authority can't be questioned without embarrassment might work better as an organizer than a planner.

You and the people who know the land and the operation well will ultimately select the best plan.

Brief the Planning Teams: Share with them the aspects of your strategic goal that influence the planning and your future landscape map. Provide an overview of the management factors and explain how they relate to developments being planned. Pass out a list of the infrastructure needed.

Explain whether you want to emphasize herding or fencing. Ask for a specific number of paddocks. Because high paddock numbers affect the positioning of water and approaches to it. But, if you have a vision of 100 paddocks, now is the time to figure out how to create them even if you cannot actually put in that many for years.

It is helpful to show some layout possibilities. If you do, be sure to emphasize the variety of possibilities rather than any particular design. Explain the herd and cell size decisions and the meaning of the planning circles. Emphasize that you're not committed to using existing facilities. However, if the country is notoriously dry, planners can't put a new water point in every paddock.

Brainstorm: Warm up with a brainstorming exercise using the game of listing solutions to some humorous problem. Then, provide each team with master maps (topography and natural features only) and planning circles. Ask them to create as many possible layouts as they can.

They should be concerned only with the layout of major features such as grazing areas (not necessarily all the subdivisions within), access to water for livestock, crop fields, working and storage facilities, landings and staging areas for timber extraction, and so on. More detailed planning (such as positioning fences and minor roads) will be done later.

You want the participants to avoid concentrating too much on any one possible layout because it may close their minds. As soon as any team has captured an idea have them set it aside and start on a new one.

Stand ready to supply information or advice. Try to offer advice in the form of a question (How will a cow get from here to there?) Stop short of outright criticism. It is important to bring a playful attitude to it and have fun.

Create One Plan Based on Existing Facilities: When all teams have exhausted their ideas, give each one a copy of the map showing the existing physical structures. Ask them to creatively develop a layout for the future based on the present situation. The existing layout is based on old concepts and is seldom the ideal for the long term. On the other hand, it deserves consideration and parts of it will usually wind up in the final plan.

Doing the initial creative planning on blank maps does not imply that the old layout is useless. You may want to use it differently by incorporating what you can of it into the final plan.

However, if you start from existing pastures, you tend to visualize those pastures as cells and think only in terms of cross-fencing them.

Evaluate the Plans: Let the planners explain how and why they did what they did. Take notes and use a marker to highlight promising concepts.

Post the various plans where everybody can see them, start discussion of good elements in others' plans.

Step 6: Designing the Ideal Plan

Those who know the ranch well should create the ideal plan from the many possibilities proposed. You are seeking the best *long-term* plan from all points of view. Boil down the options to two or three using the following processes.

Make Map Overlays Incorporating the Best Ideas: Superimpose these overlays onto a map that shows existing facilities. Think both short and long term. For example, a capital intensive layout may involve higher short-term cost but lower long-term operating costs.

Review Water Supplies: Is there enough to sustain the highest numbers in all paddocks? If not, can you afford to haul water during the times it is needed?

Review Your Checklists: Have you accounted for every item? You are specifically looking for layouts that enable you to move livestock or machinery to almost any point without constantly moving over the same ground.

Check Each Plan against the Overlays: Place your original overlays (future landscape, natural and social factors) over each possible plan and look for the good and bad features from the view of each overlay. This is the point where you will usually begin to draw up new plans based on combining the best features of the others. Let these sit a while as others mull over them while you work through the details of cost and construction. Then come back and create the final plan.

Check Your Ideal Plan against the Reality of the Land: Go out and walk the land. Compare planned fences and other developments with the reality of the topography. Adjust as needed.

Step 7: Implementing your Plan

Begin the gradual process of changing over from the old to the new. Commonly, the cost of the changeover is a major determining factor in the rate of change.

None of it should cost you outside money unless you choose to use it. The land itself should be able to generate all or most of the funds needed.

Implementing the plan gradually as the land generates the capital to do it means that you will operate at a profit year after year and your strategic goal will continue to be met.

Notes, Ideas and Practical Solutions:

Figuring Costs and Schedules: The speed with which you implement your land plan will be governed by your annual financial plan.

If energy conversion is not the weak link, building fences is seldom wise.

In many operations, product conversion will be the weak link initially—too few animals to eat the forage produced. In cases like that, available money should be diverted to livestock purchases. Then, those first years, can be used to experiment with different methods, for example, for how to make your fencing fully movable.

As soon as energy conversion is the financial weak link, develop the elements of the plan with the greatest marginal reaction.

Break the Plan into the Smallest Plausible Steps: Compute the cost of each step. The layout may involve several stages of center construction, water development and fencing. There may be some things that you will have to build at once for anything else to follow—e.g. well(s) and/or pipeline(s). These cases are costly and you would do better by finding a way to build incrementally so that every improvement pulls its full weight and will begin generating income as soon as it comes on line.

Determine a Sequence of Construction: Give priority to those things that contribute directly to producing solar dollars. A water point might increase useable range but only if you have more livestock than the currently accessible grazing area can handle.

When resource conversion is the weak link, the highest marginal reaction commonly comes from raising stock density and decreasing grazing periods. In this case, do as much fencing as you can before investing heavily in a more efficient water system.

Then there is the question of which fence to build first. Here are a couple of examples:

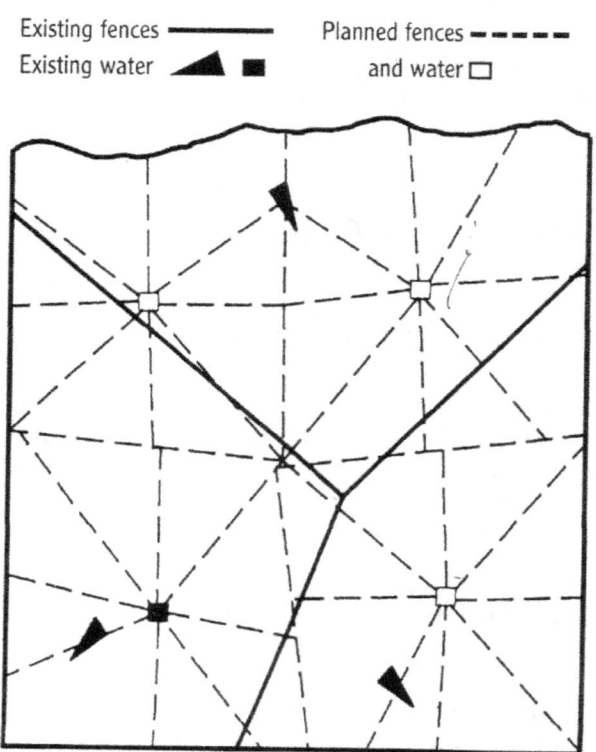

Which Facilities Would You Develop First?

Existing fences ——— Planned fences ─ ─ ─ ─
Existing water ▲ ■ and water ☐

- *Example 1: Fences or water?* (Refer to the figure below). Say your plan calls for converting three large pastures into four radial-layout cells that could be combined into one.
    o Assume developing water in the centers requires a major capital outlay. How many fences could you build before you have to change a water point to avoid a dry paddock?
    o Assume that, aside from water, the centers will contain no expensive handling facilities only consisting of a corridor and gates. Which would you build first?
    o In which center would you first develop water
    o How long would you use the existing fences before replacing them?

Which Fence First?

- *Example 2. Which fence first? The recovery approach.* (Refer to the figure below). Generally: decide which fence to build first according to how it will reduce overgrazing on the largest number of plants. (Assume that all fences cost the same and productivity is uniform across the paddock.)
    o Choice A: The two small paddocks would get all the benefit with little impact on the rest of the cell. Due to unbalanced grazing periods, plants will be overgrazed in 90% of the cell.
    o Choice B: Will affect the whole cell more evenly but no part of it intensely. Work out the grazing periods for 30 and 90 day recoveries and you will see the advantage of Choice B. It will not completely end overgrazing because 4 paddocks will not shorten grazing periods enough. But, recovery periods will be sufficient and 100% of the cell will benefit from the recovery time.

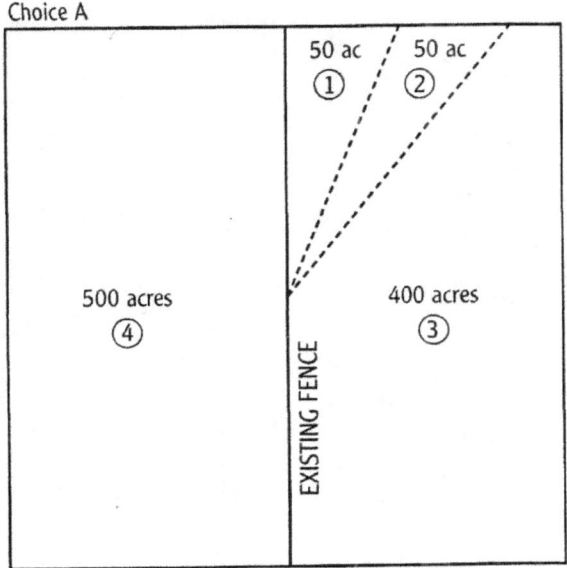

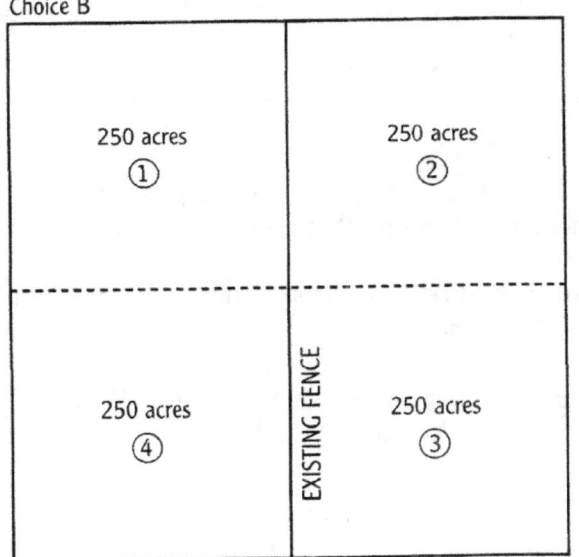

- *Example 3: Which fence first? The productivity approach.* With the productivity and fencing cost variables added, you can calculate animal-days per dollar by dividing productivity in the new paddocks by the fencing cost. This provides a good index for gaining the most production for the least amount of money. See the figure below: the plan calls for splitting each paddock in half. When the fence splits the paddocks unequally, use the productivity figures of the smaller piece.

*AD – Animal Days; ADA = Animal Days per Acre

Choosing Fences by ADA/$			
Paddock 1	Paddock 2	Paddock 3	Paddock 4
350 acre paddocks	150 acre paddocks	300 acre paddocks	200 acre paddocks
12 ADA	25 ADA	17 ADA	12 ADA
4,200 AD	3,750 AD	5,100 AD	2,400 AD
$1,500 fence cost	$2,000 fence cost	$2,300 fence cost	$1,200 fence cost
2.8 AD/$	1.85 AD/$	2.2 AD/$	2 AD/$

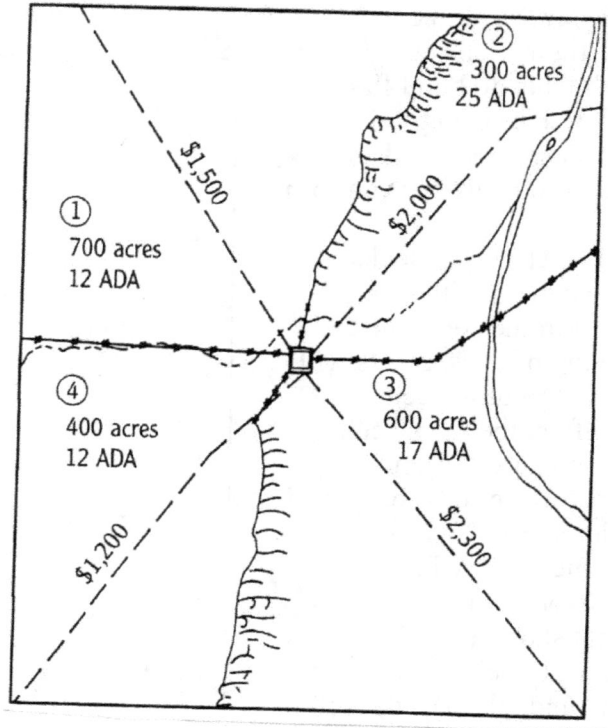

The following figure shows two possible plans for the same ranch. There are 8 existing pastures (broken lines). Subdividing them will allow an increase in density and better control over time.

Although it requires more fencing, Plan B is the less costly because it requires less costly water developments.

The land varies in productivity from 10 ADA to 38 ADA and the productivity of existing pastures varies from 1,700 AD to 28,800 AD.

Two Possible Plans for the Same Ranch

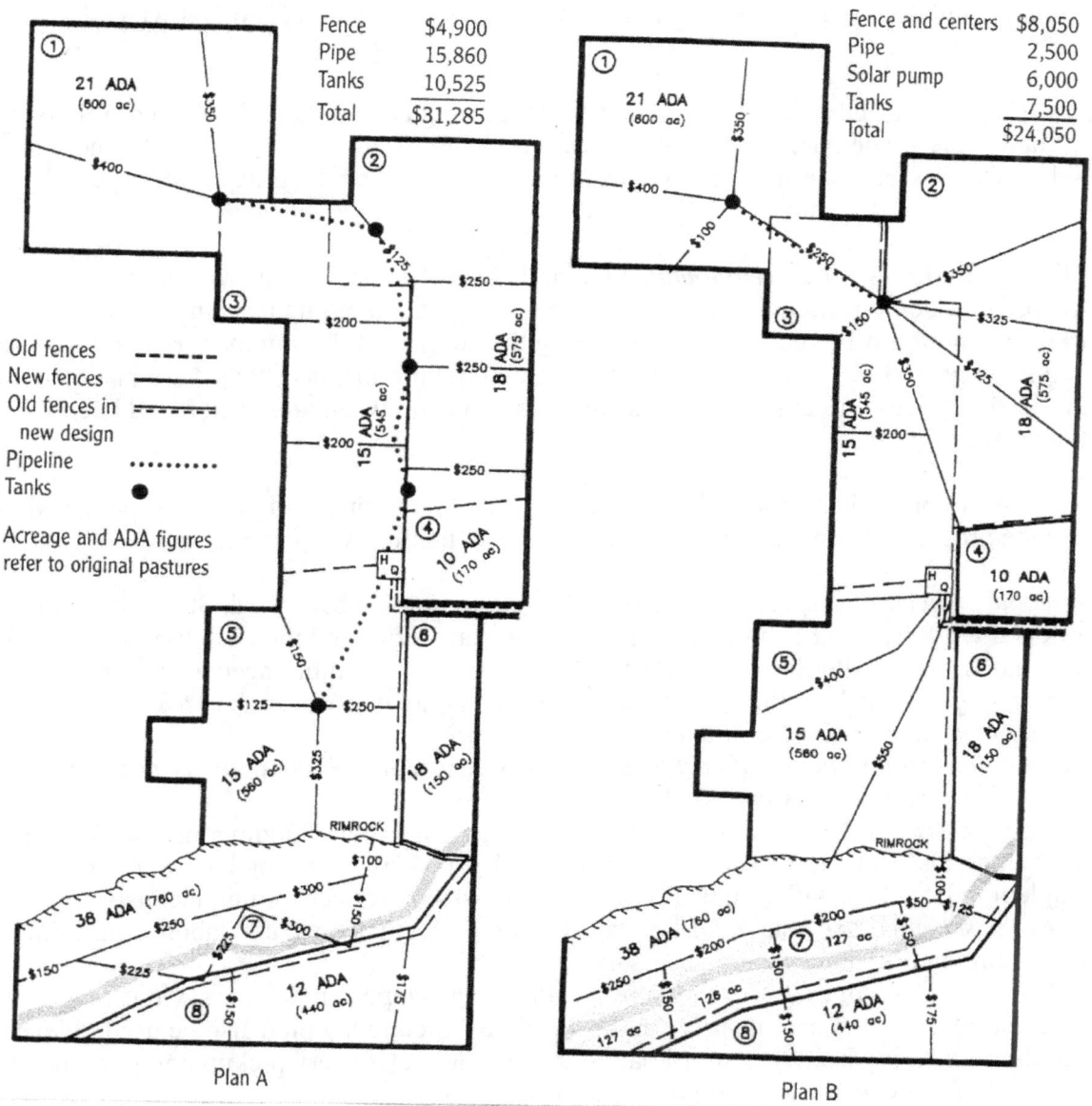

Having settled on Plan B, common sense and mathematics suggest that you should start developing fence layout by subdividing the highly productive riparian paddock 7.

AD/$ shows that fencing the shaded area gives the best marginal reaction.

38ADA X 380 (127 +126 + 127 = 380 acres) = 14,440 AD.

Fence cost = $250 + $200 + $200 + $150 = $800

14,440 ÷ $800 = 18.05 AD/$

The No Fencing Option: Permanent fencing is desirable and practical in most cases. However, there are situations where fencing of any kind is impractical or counterproductive. For example, some parts of Africa have large and diverse game populations including elephant and buffalo that wreck havoc with any type of fence.

At Savory's Holistic Management International (HMI) learning center, the land plan has relatively few "paddocks" and these are demarcated by natural features. Herders ensure that animals do not graze the same area day after day but that they do cover most of the paddock by the time they are due to move out of it.

Allotting Time and Money for Development: Loss and headache result from projects that come on line too late because of a bottleneck in the work or the financing—e.g. lambing pens that aren't ready when lambs start dropping or a water point isn't ready when the hot part of the summer arrives. Using the financial planning procedure, you are able to work the cash requirements of long-term projects into your financial projections. This will ensure that you have cash on hand when you need it and keep you aware of your debt level at any point.

Any major construction project should have its own column on the financial plan. A separate worksheet for planning the progress of the work should include at least the following steps:

1. *Break the project into separate tasks.* It is often easier to plan backward. Start with the completed work and work backwards breaking that into large categories and then subdividing each of those into smaller and smaller tasks. For example: a completed cell might need a water system, outside fences and gates and handling facilities. In turn, each one of those can be broken into tasks and sub-tasks. Note these tasks on file cards.
2. *Assign a time requirement and cost to each task.* Order pump, $800, delivery time 3 weeks; advertise for crew, $40 for two weeks.
3. *Arrange the tasks in chronological order.* Make a time line out of adding machine tape marked in 1 inch intervals, 30 feet to the year. Now lay out the file cards working backwards from the completion date. This will graphically show what tasks must occur simultaneously.
4. *Adjust the schedule.* Add slack for holidays and unexpected delays; add labor or machinery to shorten completion times where necessary.
5. *Analyze the chronology for ways to improve efficiency.* Opportunities for savings range from coordinating labor and machinery (can the backhoe rented to lay pipe also be used to level the floor in the shearing shed?) to coordinating transportation (if every pickup going to town brings back a roll of wire, can we avoid paying delivery costs?)
6. *Monitor progress.* It is easy to underestimate construction times. Reduce long tasks into *units per day* (e.g. miles of fence/day) and check the rate daily for an early warning of delays.

Layouts and Hardware

Paddock Layouts: This is the normal function of fencing but rock outcrops, flagging and other natural landmarks can be used.

Natural and ownership boundaries play a major role in the design of paddocks. There is no need to organize them around different soil or range types.

The design can reflect any number of social, managerial, political, ecological and aesthetic considerations.

A cell with paddocks radiating from a center point can't be beat for flexibility and efficiency. No other design allows for such a variety of moves from one paddock to another. However, it does not suit all situations like, for example, long narrow canyons.

You are after maximum ease of movement, minimum fencing and the most efficient use of water.

- Trails are likely to form when livestock move up or down steep slopes in the narrow end of the paddock. Sitting the center at the top of convex and the bottom of a concave slopes gives animals enough space to move back and forth along contours.
- Trails are less likely to form along fences that follow the worst terrain. Rule of thumb: build fences along ridge tops and straight down points of ridges. In hilly country, build a few of the most obvious fences first and watch how the stock move for a year or so. This will either confirm the rest of the plan or suggest changes. In steep country, fences on the diagonal up a hill encourage stock to climb the hill more willingly. Fences do not have to be straight but bends require stretch posts.
- In larger paddocks, permanent fences can be subdivided by temporary fences. Because high density causes rapid depletion of forage and intense fouling, livestock will concentrate on an area encompassed by a single moving fence almost as if a second fence were moved along behind them.

Cell Centers: Livestock will never respect fences that have no function. So, it is usually best to build cell centers before developing paddocks.

Never introduce livestock to electric fences that don't carry full current.

In highly productive pastures where existing facilities are adequate, you may want to delay building a new center. Nevertheless, design it thoroughly from the start. A few ideas are shown on the next several pages.

Cell Centers and Fence Patterns

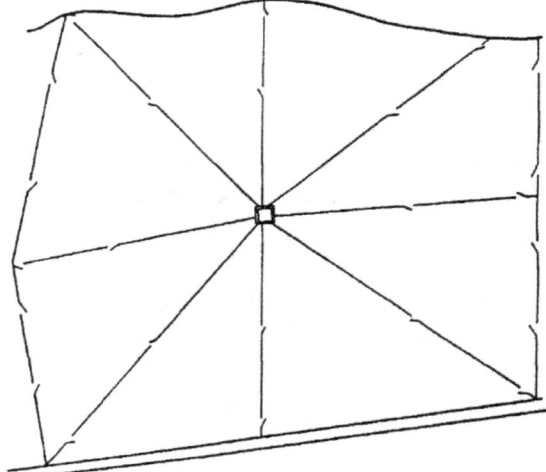

Simple radial cell with water in center. Note gates away from center

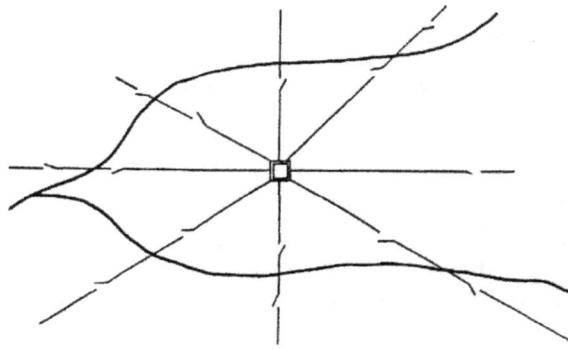

Dry center—water available in all paddocks.

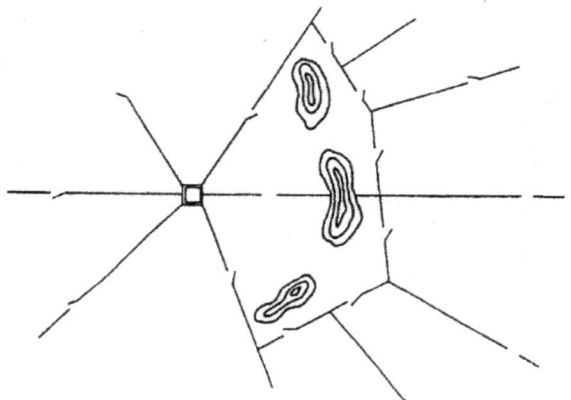

Fencing to avoid problem piece of land near center.

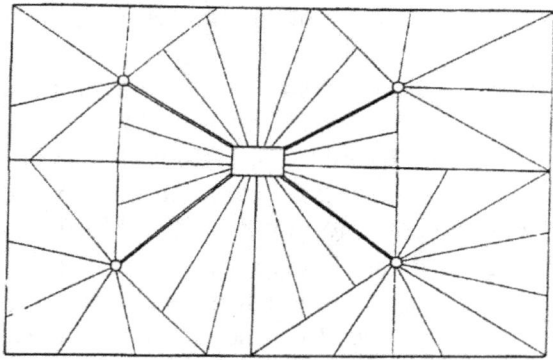

Dairy unit in center and water at satellites.

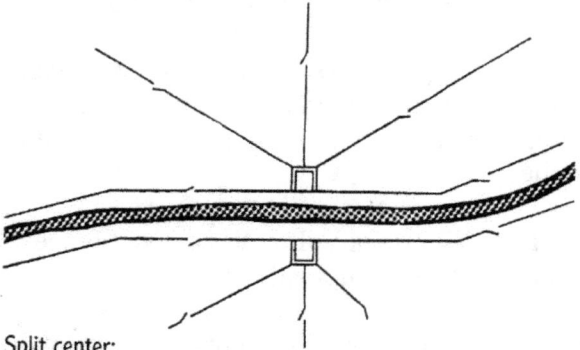

Split center:
Full facilities would be in the half serving the largest area. Half centers can be set well back from river or road.
1-, 2-, 3-, or 4-way splits are possible.

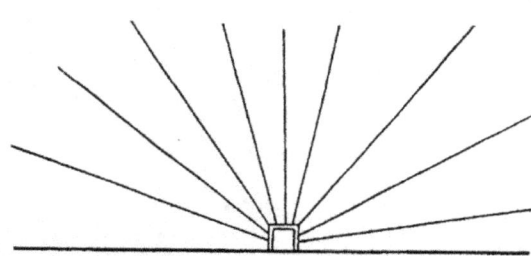

Radial fences in a fan design (takes more fencing than wheel layout).

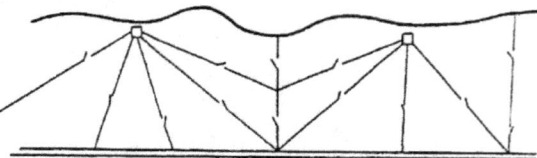

Several fans used in a narrow area.

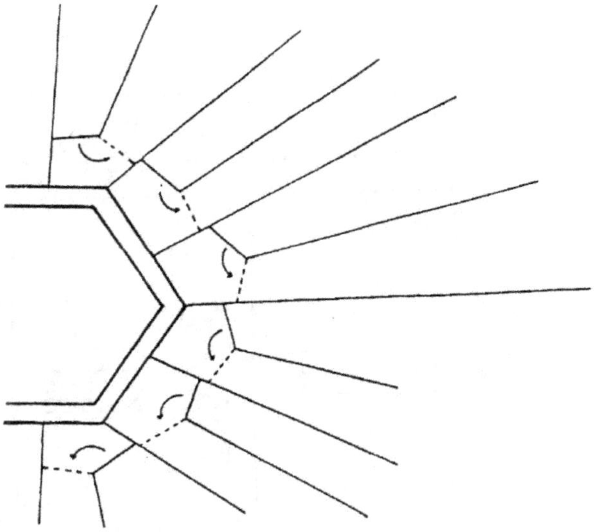

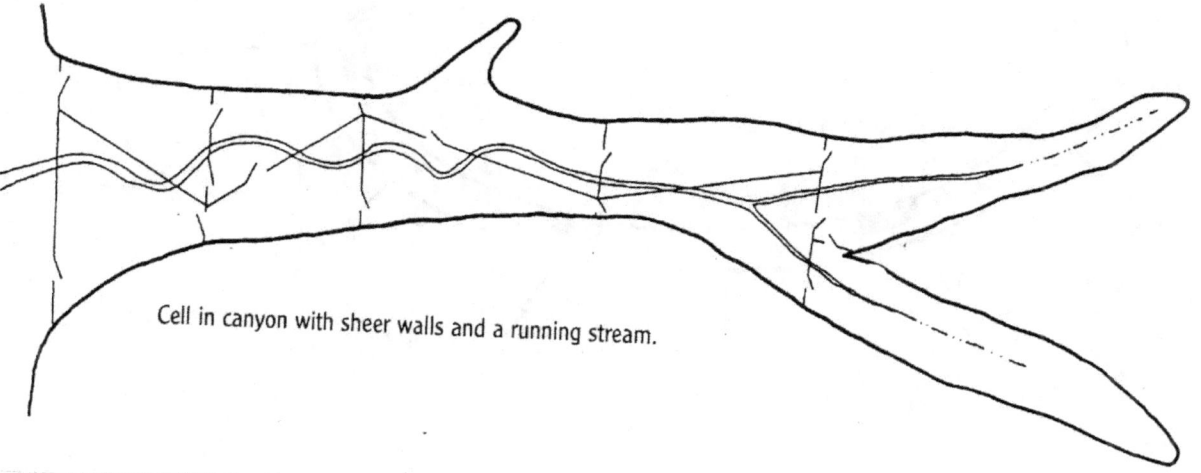

Alternate radials into a small center.

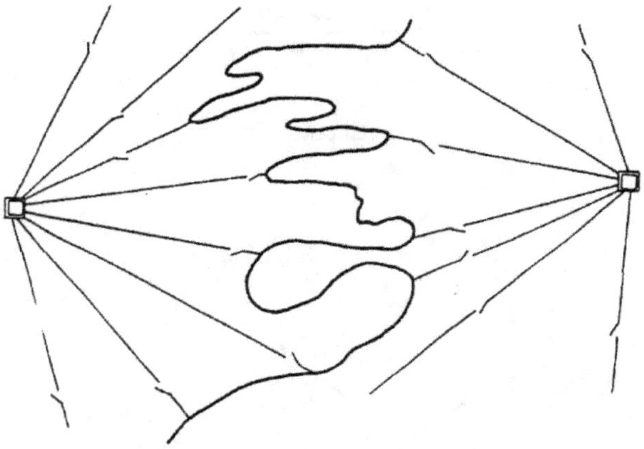

Two cells divided by rimrock.

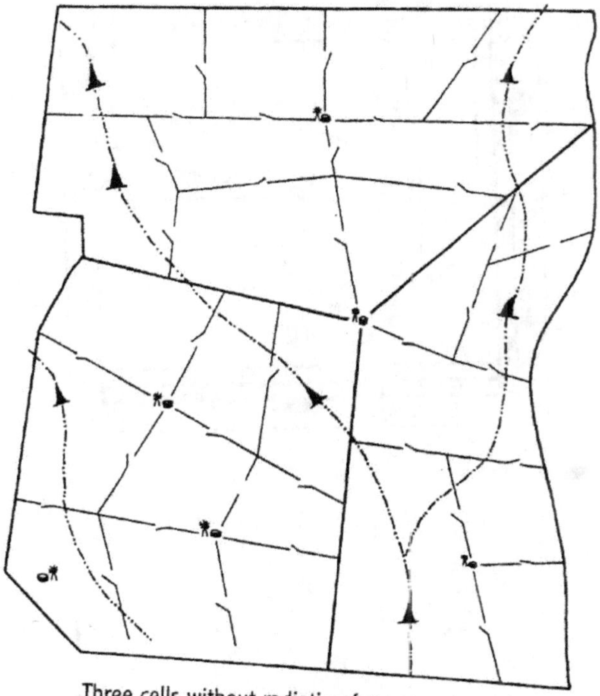

Three cells without radiating fences

Cell in canyon with sheer walls and a running stream.

Simple Cell Centers

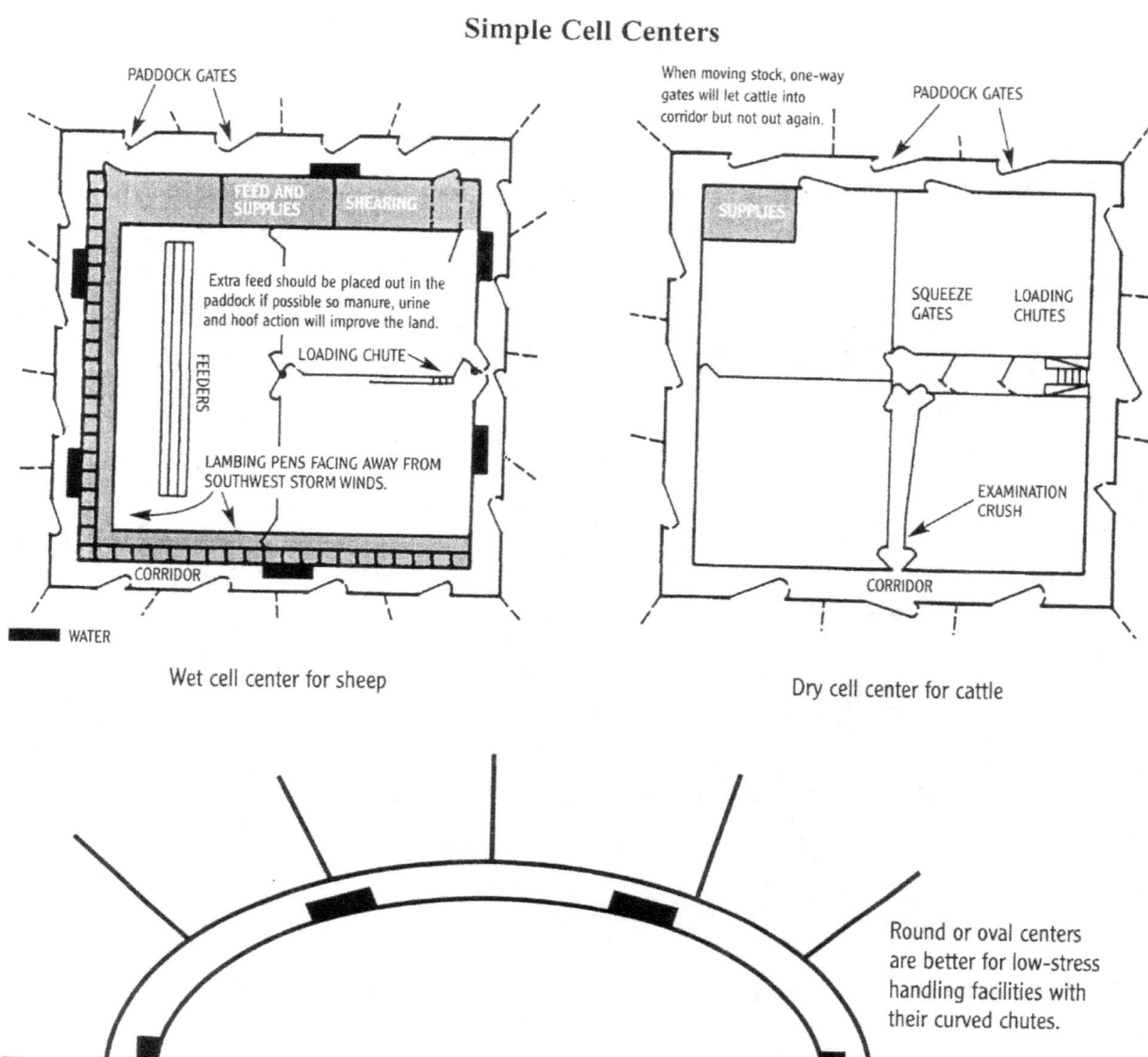

Cattle will loiter in the cell center unless you build a narrow "corridor" around the outside wide enough for a pickup. You can leave room in the middle of the center for all kinds of handling facilities. Most build the corridor first and little by little add fancy facilities inside. *Always* look ahead and leave space for future needs.

Positioning Centers: Consider access, water availability, constructions obstacles, and proximately to roads or hazards. If possible, build on higher, dryer ground.

Center Corridors: You do not want cattle lounging in the center. You do want space in the center for handling animals, storing supplies and holding horses, bulls and sick stock and a large, clean central work space. A narrow corridor around a central area will solve both problems. The corridor might even surround an entire homestead.

Figure the size of the central space according to the facilities you want in it. Be generous. Five thousand square yards will easily accommodate the daily handling of 500 to 1,000 cattle.

A fifteen feet corridor will accommodate herds of almost any size and the benefits disappear entirely at about thirty feet.

The perimeter should be long enough to accommodate plenty of gates as you eventually may have thirty or more paddocks.

Once planned on paper, lay it out on the ground with stakes and twine and test it. If vehicles have to pass through or turn around, be sure they can before you start.

Handling Facilities: Good Ideas: Curved chute layouts reduce stock stress. Temple Grandin offers free designs on her Web site (www.grandin.com/design).

Lambing Boards: Newborn lambs and kids can stay in the center until they are big enough to discourage predators. Place a board across the entrance to the paddock high enough to prevent young from jumping out but low enough to let mothers in to suckle.

Calf Traps: If you need to frequently separate mother cows from their calves (to dip them for example): A Zimbabwean rancher developed a layout that enabled one man to bring in five hundred cows and separate their calves with minimal stress. The first cows were through the dip and in the new paddock in three to five minutes. After a brief training period, there was no bawling or attempt by cows to look for calves as long as the herd was back in the paddock within an hour. (See the diagram below.) As cows with calves are moved to the gate at point A and compressed toward point B, the calves automatically duck between the poles of the trap (D) which is wide enough for them but not a cow. After a few handlings, it becomes routine. As cows move on toward C, the calves in the trap are driven through and out at point E to join their treated mothers. If the calves require treatment (dehorning, castrating, etc) this can be done within the corridor to E. Similar calf traps have been designed using the same principle—a crowding area and the chance for calves to escape backward or sideways through the bars and into the trap. Once in the trap, calves can be retained by lowering a ladder (of nylon rope or chain and rebar rods) on the inside to block the spaces. Some cows do get agitated in early training so adding a high crossbar that they can't get over might be needed.

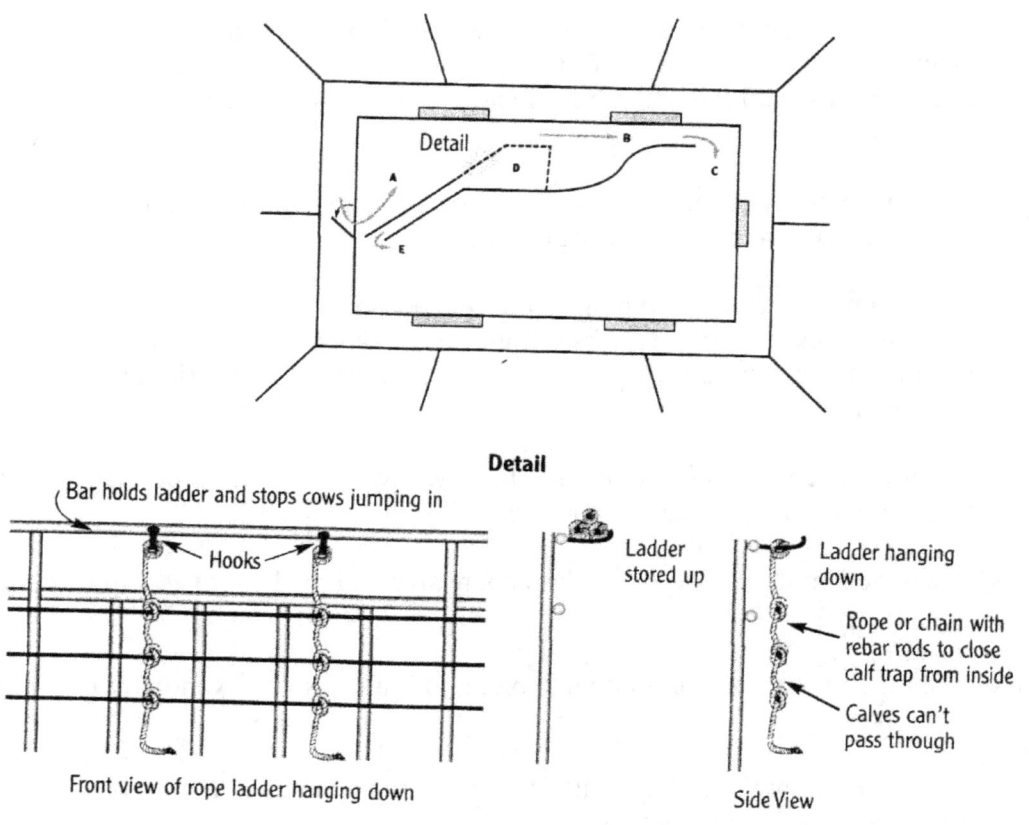

Single-Sire Breeding: This is often quoted as a reason many small herds are essential and the resulting loss in forage production is so high. Consider this layout developed in Namibia. Not only were there five hundred breeding cows in a single herd, the rancher was also able to use one bull to 90 cows. He also knew the exact date each cow conceived and was quickly made aware of any faulty bull. Here is how it worked.

Bulls were held in separate breeding pens located in the cell center with color-coded gates. Each one had a colored ear tag in each ear to match his pen color. They were fed in their pins and taken out together each day to water in a paddock well away from the cows.

Early each morning, the herders looked for cows in season that came into the center. Each cow had two colored ear tags. If they were red, she was put in the red bull pen and so on. After she was serviced, the herders let her out and noted her number and the date in a book.

Using one bull to ninety cows was possible because a bull did not service the same cow repeatedly. The date of conception was known if the cow did not return to the bull. And, if the bull was defective, it showed in repeat visits by the cows assigned to him.

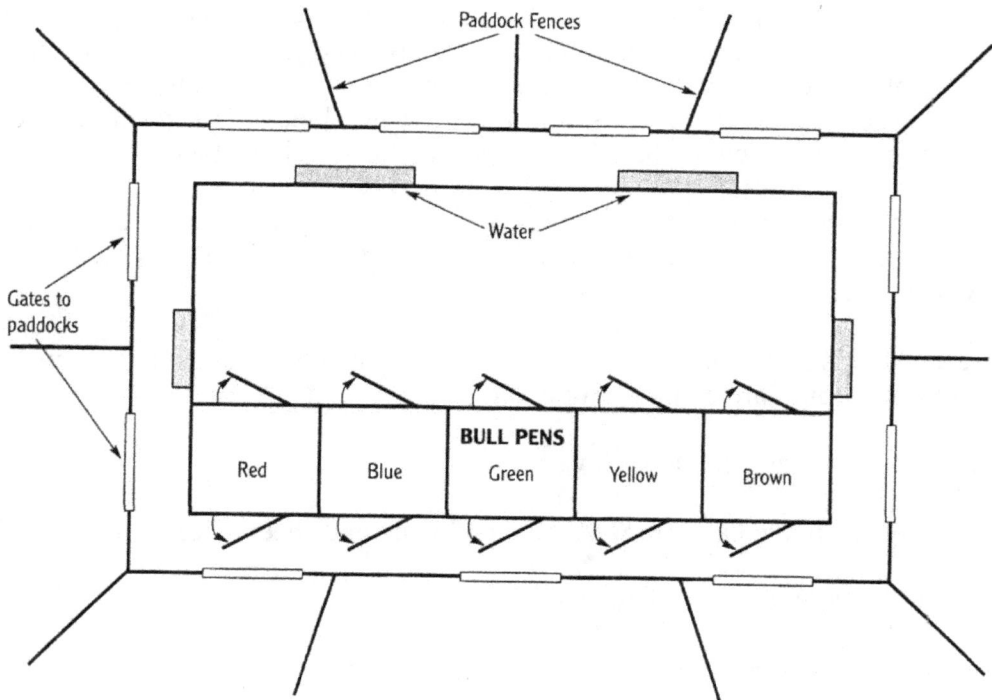

Water: If you have a stream or a dam you plan to use as water point:

- Flowing water necessitates construction of walkways to prevent over trampling and muddying the water.
- Consider fencing dams and other catchments so animals can't stand or wallow in and foul them. Permanent posts will allow you to change access areas occasionally or extend access when the level drops. Consider siphoning water to a trough.

If your water source is a well you will probably want water piped into the center. Basic principles:

- Be sure that livestock never have to wait. If all stock trust the supply, they'll come to the center only when genuinely thirsty, water and then leave. Make sure storage capacity is big enough to handle peak demand. If necessary, pipe small amounts of water from several minor sources to a reservoir at or near the cell center.
- Cattle require at least 15 gallons per animal per day.
- Rate of delivery is more important than having large troughs.
- It is wise to have several days' water stored in case of breakdowns.
- Livestock perform better with cool, shaded water than with hot water.
- Portable troughs may help spread animal impact. Use T-junctions or "tapping saddles" for take-off points that allow you to move the trough within the paddock.
- Consider water trucks with a trough on the back as a way to move livestock to distant areas.
- Long narrow troughs work better and are easier to clean. Calves can get pushed into troughs over 9 inches wide so weld bars across the top.
- Consider building ramps that allow birds and other small animals to drink without drowning. This keeps the water unpolluted while building complex communities.
- Seeps or pools are easy to construct for shy creatures that won't brave cell center fences to get water.

Fences: Free-running slick wire fences electrified by high-voltage New Zealand-type energizers are the most cost efficient and least damaging to wildlife. Posts do not need to be as sturdy as conventional fencing. It is the fear of shock and not the strength that is important. Should a fence be put in the wrong place, it is not a big problem or costly to reposition. It is not a good idea to use electric fencing where crowded animals are forced to touch it when they get pushed (as in cell centers).

How Many Wires? On dry soil where grounding is poor, two wires (one hot and one grounded) will contain almost any domestic stock that are not seriously stressed and are well trained. Where grounding is good, one wire will suffice.

Spacing the Wires: There is no ideal. Two wires 3 inches apart and between 24 and 30 inches above ground (top hot-bottom ground) are common for cattle.

For smaller stock the bottom wire should be lower.

Three wires may be needed in the case of mixed herds of cattle and sheep with the bottom wire hot.

The same fencing will work on the cell perimeter.

Power can be switched off to all paddocks except for those containing livestock.

Posts: Almost any wood, steel, or fiberglass post will work. Poor-quality fiberglass decays in sunlight so ask for references.

Fiberglass and lighter wooden posts will slide up and down in their holes but will drop back into place.

If the wire bends around a tree, run it through plastic insulator hose. Tension will hold it in place.

Attaching the Wires: The wire must run free (slide unrestricted) and attachments must be strong, durable and easy to install. Free running wire distributes force through the whole fence to end strainers designed to take it.

Repairing breaks will create knots that will cause problems when winding up the wire. So, use eyes that allow you to lift out the wire and drop it to the ground before you wind it up.

Straining Posts: Substantial posts will be needed at each end of the fence, at significant bends and along a long length of fence. But most of the posts only hold up the wire.

For load bearing (stretch) posts, set the post deep (as much below ground as above) with an 8 to 10 ft. diagonal brace. The pad at the end of the diagonal can be rock or treated wood.

Generally, a single stout wooden post can handle strain in the middle of a fence because the pull on opposite sides cancels out.

Grounding: Poor grounding is one of the most common faults. Three or four 6-foot steel rods driven into the ground and attached in series will usually solve the problem.

It is common to locate the charger at the cell center and distribute power to the fences through wire strung on posts above the paddock gates. If the ground is above the hot wire, it also serves as a lightening deflector.

Gates: For the sake of flexibility, you can't have too many gates. You must be able to move livestock from one paddock to another at several points along the center.

Any kind of gate will work.

When you have a two strand fence and your ground is level enough to allow a good distance between posts, you can simply push the wire to the ground and hold it down with a couple of buried hooks while stock pass over it.

Gully and Stream Crossings: Pass the wires straight over the gully or crossing. Then attach wires or small chains alternately to the hot and cold wires. These chains then dangle down into the gap and prevent stock from passing but allow water and debris to wash through.

To keep the drop wires or chains from shorting, run the fence wires side by side on opposite sides of wooden posts then hang wires or chains every foot or so on opposite wires.

Educating Livestock: Animals usually learn quickly. But, it may help to tempt an unruly bunch of goats with aluminum cans smeared with molasses and clamped to the hot wire. It works best on a wet day or when the cans are hung so the animals are likely to touch the ground.

If you annually acquire a great number of untrained stock, start them out in a "training paddock" at a high enough density to be sure a good number get zapped. Adding an extra wire or two, hosing down the ground and retaining a few experienced lead animals will speed the education.

Unshorn angora goats and sheep that have never tasted molasses may learn that molasses stings but nothing about fences. Wet ground may help them get the point.

Training stock to move to a signal makes fencing more effective. It is also useful for concentrating livestock to produce herd effect.

Training ground rules:

- Choose a unique signal that they only hear when you intend to move them.
- A pick-up horn can cause problems, for example, when a passing vehicle honks.
- Always associate the sound with a reward. Make the message absolutely clear while you are training them.
- Don't mix reward and punishment. If you drive them while blowing a whistle, they will never get the point.
- Consider taking a training course (like Bud Williams') in low-stress animal handling.
- Mix older, trained animals with newcomers.
- Start the training in smaller paddocks where animals can see each other.

(From: David W. Pratt. *Training Livestock to Electric Fences.* Livestock & Range Report No. 925 Fall, 1992. Napa & Solano Counties, CA. Livestock/Range Management Program

Whether building permanent fences with high tensile steel wire or temporary electric fences with polywire, an electric fence is not finished until animals have been trained to respect it.

The training area should be a small paddock. Keeping the area small will reduce the time it takes animals to learn about the fence. It will also minimize the time needed to gather and return the animals that get out during training and reduce the time required to build and mend the training fence.

When you turn stock into the training area, keep an eye on the animals but leave them alone to discover the fence on their own. Stock are curious and will investigate the fence. As they do, they'll get their first lesson. When first shocked, animals don't know how to react. Some back up. Others bolt ahead and may go through the fence. When stock get out, gather them up and put them back in. If the training fence was built using polytape or polywire, you may need to fix the fence.

When an animal investigates the fence a second time, it usually does so prepared to back up. I have never seen an animal challenge a fence a third time unless forced to do so. If an animal continues to challenge the fence, cull the animal.

Depending on the number of animals and the size of the paddock, training usually takes no more than one day.

Some people put hay or grain across the fence to give stock some incentive to cross the fence. This can increase the speed of training but is usually unnecessary.

Do not herd animals into the fence. Stock need an escape route. If crowded into the fence, animals may have no choice but to go through the fence.

Sheep are the most difficult class of livestock to train. Wool is an effective insulator, and therefore sheep are best trained just after shearing. Some producers have trained sheep by attaching cut out aluminum cans containing a little molasses to the fence. When sheep come up to lick the can, they get shocked and learn quickly to respect the fence. Make sure the cans do not touch ground wires!

Every time I have tried to control cattle or sheep with minimal electric fences (1 to 3 wires) without first training the stock, I have had to spend hours gathering stock and mending fences. When I have taken the time and effort to train the stock, the fences have been effective. If minimal electric fences are to consistently hold livestock, training is essential!

Lesson 7: Grazing Planning

General Principles of Planned Grazing

The traditional goal of "producing meat, milk or fiber" is a byproduct of creating a landscape and harvesting sunlight. To harvest maximum sunlight, you must decrease the amount of bare ground. To create such a landscape, livestock movements must be planned. You must also plan for wildlife, fire, drought and other uses.

The factors influencing your plan are recorded on a chart:

- When you breed and wean
- When and where areas will be covered in snow or threatened by fire
- When and where antelope are dropping their young and ground-nesting birds are laying
- And so forth.

A good plan can deploy livestock to:

- Reduce or eliminate problem plants including brush
- Heal gullies
- Maintain wildlife habitat
- Decrease grasshopper breeding sites
- Handle unexpected fires, flash floods, droughts and other catastrophes.
- And, at the same time, produce high volume and quality forage to provide the best possible plane of nutrition with the least possible supplemental feed for livestock.

The Planning Approach

Many farmers and ranchers offer vigorous arguments for why planning is impossible—"Too many things change all the time." But that is exactly why we need to plan. We plan because we cannot be sure of what will happen next. The main benefit of planning is peace of mind. Panic and loss of focus in emergencies can destroy you. If you have planned well, you can truly relax in the most alarming of situations.

Planning is a *process* that incorporates modifications based on continuous feedback.

There are many parallels between agriculture and the military. Generals, like farmers, must not only know how to plan but also how to re-plan instantly if all fails. The planning process described herein is based on military procedures developed to enable humans to handle many variables in a constantly changing and stressful environment. The method adapts this simple planning procedure to biological use because the same approach is just as effective in managing complex land, wildlife and livestock situations.

The Checklist

There are many factors that impact any plan and they all cannot be addressed at once.

The checklist is a guide that prevents problems in the decision making process but it is more than just a checklist. It is a sequence for making decisions that take into account the effect of one decision on another. The sequence of questions is in a specific order so that the answers build upon one another. Since your mind can only handle one thing at a time, you record each step on a grazing planning chart, wipe that completely from you mind and move on to the next step.

The overall plan that will emerge will cover every imaginable detail and will be the best plan possible for the present.

Grazing planning ties in closely with long-term infrastructure planning discussed in the lesson that immediately precedes this one.

The Planning Chart

You record the details of each step in the checklist on the planning chart. The chart is divided into 7 month sections for the sake of size. The rows represent paddocks and across the top are smaller divisions down to the day.

Problems and special considerations can be shown with color codes: For example,

- Orange in paddock 3 in May = poisonous plant problem
- Brown in paddock 4 in August = lack of water, etc.

Then, livestock moves are planned within the context of all these factors using the slowest moves that stock are likely to make in periods of slow plant growth. The emphasis is on *recovery* periods rather than *grazing* periods (recovery periods show up best when plotted on a chart).

After that, moves are plotted *backwards*. In other words: First, if, for example, certain areas are reserved for animals at crucial times. Then you indicate where the animals would have to come *from* to get there, and so on, backwards.

There is space at the bottom of the chart for planning figures and calculations. Planned and actual animal days per acre during each grazing period in each paddock are shown in the upper left of the main body of the chart. This reveals how much, on average, every acre of ground is required to contribute to total forage consumed. In turn, this allows for fine-tuning future plans and increasing the accuracy of assessments of forage availability.

When to Plan

You should plan and monitor continuously, control deviations as soon as possible and re-plan when necessary. But, major grazing planning is done twice a year.

The first time is at least a month before the main growing season begins. You are trying to grow as much forage as possible and do not have to plan to a specific date because you do not know when growth will slow or end or exactly how much forage will grow before that date. Thus, this is an "open ended" plan.

Plan the second time toward the end of the growing season, when forage reserves available for the non-growth period become known. The purpose of this "closed ended" plan is to ration out the forage over the months ahead to a theoretical end a month *after* your most pessimistic estimate of when new growth could occur—this is your *drought reserve*. Where rainfall is very low and unreliable, there may be an overlap in the reserve in the closed plan and the start of the next open plan due to the necessity of making drought reserves extend to as long as a year or more.

Record Keeping

Actual grazing times, animal days per acre, weather and growth rate information kept on the chart provide information for future plans. However, you should not turn the chart into nothing but a record keeping device. In fact, most of the record keeping done on most ranches provide little return for the time and effort put into it.

A grazing chart without the checklist is a waste of paper.

Assessing Stocking Rates

Stocking rates should not be based on the old belief that overgrazing is a function of animal numbers. Instead, overgrazing reflects timing. So, stocking rate is determined by the actual volume of forage, the time it must last and the strategic goal.

If the number of ADA of forage a herd will need for a whole season can be calculated, then the number of animals a single acre must feed for one day can also be calculated as can the area of land needed to provide forage for 1 animal for 1 day. For example:

Assumptions: Size of the ranch: 15,000 acres; Size of the herd: 950 animals; Dormant period: 180 days; Plus Drought reserve of 40 days = 220 days

To calculate animal days per acre (ADA) of forage needed: (950 animals X 220 days)/15,000 acres = 13.9 ADA. In other words, a square of 1 acre must provide enough forage for 13.9 animals today. 1 acre = 4,840 sq yards. $\sqrt{4,840}$ = a square of 69.57 yards on each side.

To calculate the area of land needed to provide forage for 1 animal today: (4,840 sq yards/acre)/(13.9 animals/acre) = 348.2 sq yards/animal. Thus, $\sqrt{348.2}$ = an area of 18.6 X 18.6 yards must feed 1 animal for 1 day.

So, you need only pace off a square about 19 yards on a side and judge whether or not it would feed one of your animals for one day.

The method's weakness is the human tendency to fudge in selecting and judging samples. A random sample should include all areas including roads, hillsides, brush, etc. If your sampling procedure does not include those areas, you should subtract the amount of land tied up in these areas from the total acreage of the ranch.

Judge each sample conservatively. Estimate like a cow by imagining filling a large bag in eight hours using only one hand to pick the material. If you imagine any difficulty in doing this, a cow would have trouble too. If you hesitate even for a moment, fail the sample.

Academics insist that the forage be clipped and weighed. We reject this because:

- Livestock do not clip grass. What they do more resembles harvesting by the handful while avoiding old stems.
- Ranchers can make good estimates by the eye "like a cow."
- Clipping and weighing takes time and thus reduces the feasible number of samples. More simple judgments yield better overall accuracy.

You can vary the one animal per day question to suit specific needs. For example, if you are dealing with lactating cows ask yourself, would this sample feed one cow *very well* for one day? Or, is it enough for a 6 Cwt steer? Or, what will a pronghorn be able to find after the cows are moved tomorrow?

The same technique will work during the growing season although re-growth makes it tricky. If the forage in a paddock is depleted before the plan calls for moving the livestock, the stocking rate may be too high or the grazing period too long. You could cut grazing days from a poor paddock and add them to a better one. But remember, shortened grazing periods add up to a much shorter recovery period which, in slow growth periods, ensures overgrazing.

Two things determine the foraged taken out of a paddock: the number of animals and the time they spend in the paddock. So, if you run out of forage before the animals are due to leave, you either have too many animals or have allotted too many days. By minimizing time, you leave only one variable to judge—animal numbers. If you can't reduce the number of gazing days without overgrazing because the recovery period is too sort, then you may be overstocked. If the problem shows up in only one paddock, you probably just misjudged the paddock. However, if it shows up in several paddocks, you are almost certainly overstocked.

Stocking rates can be checked any time of the year. Suppose you check in the middle of a very dry summer and discover that the next several paddocks you are due to enter do not have enough forage for even the shortest feasible grazing period. You would immediately re-plan.

Basing Stocking Rates on Annual Rainfall: Some have suggested using running averages of precipitation to determine stocking rates. But, this is doomed to failure because averages do not reflect rainfall *effectiveness* which depends on many factors. The only practical way of determining stocking rates is to base them on the actual forage produced every season which is, in turn, determined by the effectiveness of the water cycle.

Conflicts with Wildlife: Grazing planning can help overcome conflicts between livestock and wildlife. It gives positive control over where livestock will be on any given day while taking into account wildlife needs for nesting sites, cover, etc. Furthermore, the concentration of the herds means that most areas do not have livestock on them at all for up to 90 plus percent of the time.

A method for accurately assessing wild animal numbers is an area that needs a great deal of research. The primary difficulty is assessing the numbers of wild animals reliably enough to use them in planning. We can only make educated guesses based on the animals' condition and evidence of forage use throughout the year. While over-browsing by wildlife is relatively easy to observe, overgrazing is not. Over-browsing leads to high juvenile mortality (they can't reach as high as adults) but overgrazing does not (even the youngest can reach the ground).

Planning without Paper

Millions of acres are occupied by people who subsist by grazing livestock. Most of these lands are communally owned and are located in brittle environments. The inhabitants are unlikely to have the ability to plan on paper. But, they do have good memories, are observant, are knowledgeable about their animals and the land and are able to herd their livestock.

These common sense attributes plus a clearly thought out strategic goal (in which they have ownership) renders such people perfectly capable of working out the detail of the plan.

Monitoring and Controlling

A plan serves very little purpose unless its implementation is monitored and deviations controlled. In simple situations, a rough plan in the head with common sense adjustments as you go along may be enough. But unfortunately, such simplicity is nonexistent in the management of biological resources. Progress toward the future landscape must be monitored using the biological monitoring process described in another lesson. What we are referring to here and now is monitoring the grazing plan itself.

Monitoring Daily Growth Rates of Plants: If you have a small number of paddocks (fewer than 100), where grazing periods run from three to 15 or more days, *monitor daily growth rates throughout the growing season. Always plan for slow growth but when rapid growth occurs, shorten grazing and recovery periods or you will overgraze.*

Keep in mind that plants can be overgrazed in three ways:

- When exposed to animals for too many days.
- When animals move away but return too soon.
- Immediately after breaking dormancy when new leaf is being produced from stored energy.

The greatest damage is normally done when shortened grazing and recovery periods are not lengthened after growth has slowed back down. The shorter recovery periods coupled with slower growth mean that animals are returned to paddocks that have not yet recovered from the previous grazing.

Growth rates should be judged on bunch grass ranges based on severely grazed individual plants and not the overall view. As soon as livestock leave a paddock, find and mark several severely grazed plants that are alongside un-grazed plants of the same species. The plants can be marked with a 2 to 3 ft. piece of wire and brightly colored engineer tape so they are easy to find. A kink can be put in the wire to indicate the initial height of the grazed plant. Also, the un-grazed plant can be used as a yardstick.

If, weeks later, the plants have barely re-grown and livestock are due back in that paddock, you know your movements are too fast for the currently prevailing slow growth rate.

On the other hand, if the plants are growing so fast as to be barely distinguishable from the un-grazed ones and livestock are not due back in the paddock for a month or more, your movements are too slow. Overgrazing is likely to have occurred in the paddocks grazed since you marked the plants—that is unless you have very high paddock numbers and, as a result, always short grazing periods. Also, where you have few paddocks and long grazing periods, livestock performance will suffer.

Exactly when is a plant recovered? Actually, too many factors obscure the question for there to be a definitive answer. The best we can hope to do is observe severely grazed plants and consider them recovered when they resemble the un-gazed plants alongside them under identical conditions.

In planted pastures or natural grasslands in less brittle environments, individual plants may not be distinguishable. Daily growth rate can be monitored by examining the color of the plants (darker green implies faster growth) and the height of the vegetation both in front of and behind the livestock.

The general rule is fast growth = fast moves and slow growth = slow moves. But, what if some species are growing fast and other species are growing slow, as may be the case where you have both warm and cool season species. (See the figure on the following page.)

Assume you have only 5 paddocks so that graze periods have to be long no matter the growth rate (about 7 to 22 days if you use recovery periods of about 30 to 90 days). Suppose at the end of April (Point A) you detect slow growth on severely grazed warm-season grasses and fast growth on severely grazed cool-season grasses.

If you move fast (7 day grazed period), you are doing the best you can with 5 paddocks for both species during the grazed period but, what about recovery? Look at the chart. The cool season grasses are likely to be dormant when you return in 30 days (Point B) but the warm season grasses could be growing fast. If they continue to grow slow, you are likely to overgraze the ones that were severely bitten.

Your alternative is to move slow (22 day grazed period). But, if you do, you increase the chances of over-grazing some of the cool season plants that were growing fast.

But look at the livestock management year (top row). The cows are lactating in late April and the bulls will be joining them on 1 May. So, moving faster for better animal performance might justify the risk of overgrazing the warm season species if growth continues to be slow into June.

Actually, this situation is not nearly as serious as it sounds provided you are applying high animal impact and your planning is good. You will not overgraze nearly to the degree that we accept daily under conventional range management.

Also, *this furthers the case for more paddocks.* The situation would be quite different with 35 or more paddocks. Recovery periods of 30 to 90 days mean grazing periods of only 1 to 3 days. Even the fastest growing plant will not be overgrazed during the longest grazed period (3 days).

Holding stock in paddocks too long could lower conception rates. So, moving faster (1 day grazed period) might improve conception rates and weaning weights. But, on the other hand, the shortened recovery rate (30 days) and high stock density might result in overgrazing of warm-season plants if growth continues to be slow.

So what is the take home? In any case, the land will forgive a reasonable compromise as long as you have a good grazing plan and keep applying high animal impact.

Furthermore, you overcome the dilemma of having to cope with simultaneous fast and slow growth with high (100 or more) paddock numbers. The high stock density and short exposure time of plants to animals keeps the proportion of leaf-to-fiber in the plants high. This, along with short grazing periods, means better animal performance on longer rested grass.

The Importance of Monitoring Actual Growth Rates: Do not make assumptions about growth rates just because it rains or the temperature is right.

When monitoring of growth rates tells you plants are growing rapidly, it is time to exert control. If you only have a few paddocks where the planning will call for two grazing periods—one for rapid growth and one for slow growth—switch to the shorter grazing periods.

Then, keep monitoring and, as growth rates slow down, lengthen the livestock moves. As you monitor and adjust, keep an eye on the recovery periods.

Never graze on an average because growth rates are seldom average. They swing widely between fast and slow through the growing season especially in areas where precipitation is erratic and water cycles are non-effective.

Re-planning: Rarely does this become necessary. However, on occasion, a prolonged period of rapid growth will keep grazing periods short for so long that the plan gets skewed and animals scheduled to be in a certain place will get there sooner than planned. All you need to do is erase the plotted grazings over the next few months and re-plot them.

Other situations where re-planning would be needed might be after a major catastrophe like a major fire or breakdown in the water supply.

Droughts seldom require re-planning because it should have been planned for from the beginning.

Review and More Basics

A cell is a piece of land on which livestock moves are planned as a unit and recorded on a single grazing chart—i.e. it is defined by planning for livestock grazing and *not* by the layout of fences.

Plans can never be perfect, especially at first. Since you must balance all the social, economic and environmental factors, you may have to settle for some overgrazing and partial rest until you can develop the necessary infrastructure—fences, water points, etc. Obviously, the timing of these will be determined by your financial plan. However, the planning process will coordinate acquisition of livestock and infrastructure so that you remain profitable at each step along the way.

Of the tools we have for managing land, the primary ones used in grazing planning are: rest, grazing and animal impact. In turn, four management guidelines govern the use of these tools: (1) Population management (stocking rate); (2) Time; (3) Stock Density and (4) Herd effect

Why Plan Grazing? The excuses most commonly offered by ranchers for why they do not plan are, in fact, the reasons why they should. If you knew exactly what the weather and all the many other many variables were going to do, you wouldn't need to plan—provided you could keep all that in your head.

Measuring Forage Utilization in Days

Animal Unit Months (AUMs) work well for calculating grazing fees. But, Animal Days (AD) work much better for calculating grazing taken in a single grazing period or over a season.

Animal-Days: An animal-day is a measure of forage quantity—i.e. the amount an animal eats in a day. Roughly speaking, it is enough forage to fill an animal's stomach. With experience, you will learn to refine your estimates to include the amount of forage trampled as well as that used by wildlife.

Factors for Conversion into "Standard" Animal Units: There are several approaches to this but the following is the one we will be using:

A cow = 1 animal unit
A cow and her calf = 1.5 animal units
A weaned calf = 0.75 animal units
A bull = 2.0 animals units
A 500 lb calf = 0.5 (50%) of an animal unit
A 750 lb steer = 0.75 (75%) of an animal unit
A 2,000 lb Bull = 2 (200%) of an animal unit
5 Adult sheep, goats or pigs = 1 animal unit
10 lambs, kids, or piglets = 1 animal unit

If you run more than one class of animal, you will have to adjust your animal day figures accordingly. You can translate each class of animal numbers into standard units that reflect the kind of animal and its size. Then add all the figures to give you one animal-day figure. For example: Say your herd is composed of cows, bulls, calves, ewes and lambs.

100 Cows	=	100 AUs
4 Bulls	=	8 AUs (4X2)
85 calves	=	43 AUs (85X0.5)
200 ewes	=	40 AUs (200 divided by 5)
300 lambs	=	30 AUs (300 divided by 10)
Total	=	221 Animal Units

So, if this herd spent 5 days in a paddock, they would consume 1,105 (221 X 5) animal-days of forage.

These standard units do not consider the variances in the physiological needs of the animal (due to gestation, lactation, etc.). But, for all practical purposes, a lactating cow will not consume much more than a dry cow. Her nutritional requirements are, in fact, greater but she, just like the dry cow, will eat until she gets full. So, sophisticated tables that factor in such things offer no real advantage over simple field checks.

Animal-Days per Acre (ADA): Although herds of different sizes may spend different lengths of time in paddocks of different sizes, you can still estimate how much forage the *average* acre in that paddock will supply.

Animal-days per acre = $\dfrac{\text{animals X days}}{\text{Aces of land}}$

For example: 50 animals in a 100 acre paddock for 4 days will take 2 ADA [(50 X 4)/ 100]. On the other hand, 50 animals in a 50 acre paddock for 4 days will take 4 ADA [(50 X 4)/50]. So, what does all this mean in practice?

Suppose you have a 500 acre paddock that a herd of 100 head have been through three times this season already—once for 3 days, once for 4 days and once again for 3 days for a total of 10 days.

$\dfrac{100 \text{ animals X 10 days}}{500 \text{ acres}}$ = 2 ADA   As compared to:

Another paddock of 100 acres in which the same herd spent one day each for the three cycles (a total of 3 days) and then, on the last day, they were joined by another 100 head that had been gathered for shipping.

$\dfrac{(100 \text{ animals X 3 days}) + (100 \text{ animals X 1 day})}{100 \text{ acres}}$ = 4 ADA

Such comparisons can allow you to change your plans in subsequent years if some paddocks seem to respond differently than others.

Grazing, Overgrazing, and Growth Rates

Overgrazing does its greatest damage when plants are green and growing. It occurs under two conditions: 1) When animals remain in the same area for so long that they repeatedly re-graze the same plants; or 2) When they return to an area and re-graze plants that have not had time to recover from the previous grazing period.

A grazed perennial must use stored energy to grow new leaf. The more leaf removed, the more energy has to be drawn from the roots for the plant to recover.

Contrary to traditional belief, annual grass plants *can* be overgrazed. But fortunately, this is usually avoided when recovery (and thus grazing) periods are based on the requirements of perennial grass plants for recovery.

Recovery Requirements: Runner type grasses tend to recover quicker than bunch grasses because a smaller percentage of their leaves are removed in a single grazing of a given height. Irrigated runner grasses that are rapidly growing can recover in as little as ten days.

On the other hand, bunch grasses under erratic growth or long periods of little or no growth can take 90 days to a year to recover. Typically, recovery is rapid in low-lying humid areas but slower at high-elevations and in arid areas.

When judging recovery times, it is best to use your own experience on your own place rather than some "suggested" time period.

Generally grasses will have recovered enough when they look like the nearby un-grazed plants. You can judge this by placing a flag near (or cage over) severely grazed plants and monitor their actual growth.

There are different degrees of recovery. For example,

A bunchgrass grazed severely once in the growing season and once again in the dormant season may not be overgrazed but will not develop as extensive a root system as it would if grazed only once during the year. Operators more concerned with high animal production and want to avoid overgrazing would opt for grazing the bunchgrass twice. Others who want to achieve maximum root development would opt for a single grazing and a greater degree of recovery.

If your goal is to achieve a complex, stable grassland, *adjust the timing to the most severely grazed grass plant and never to the average.*

If you want to avoid any overgrazing at all, you must have enough paddocks to ensure that grazing periods are 3 days or less in any one of them. For example, if your growing season is 180 days and you use that as your recovery period and you are running one herd through 100 paddocks, each paddock would be grazed (on average) 1.8 days and you would probably *not* overgraze any grass plants.

When Recovery Rates Vary: You may have different grasses that recover at different rates. If you only have a few paddocks (and therefore have to use long grazing periods in order to maintain necessary recovery periods) you will probably overgraze some plants. Rapid moves favor the fast growers but return the herd too soon for the slow ones. But, partial rest is far more damaging than overgrazing some plants.

To eliminate the problem, *build up paddock numbers as quickly as your financial situation will allow.* Either that *or herd the animals if the costs pencil out.*

Grazing Periods and Recovery Periods: Grazing periods are linked to recovery periods. *When you shorten the grazing period in one paddock, you shorten the recovery period in all paddocks. So, plan the recovery period first and let this dictate your average grazing periods.*

Mathematically, given equal paddocks, the grazing period (GP) is the desired recovery period (RP) divided by the number of *other* paddocks—e.g. GP = RP/(Total paddocks − 1). For example: Say you have a cell with 6 paddocks and you desire a 90 day recovery period in periods of slow growth: 90/(6-1) = 90/5 = 18 days average GP.

Then say that, during rapid growth, plants need only 30 days to recover. In that case, the average grazing period would drop to 6 days. 30/(6-1) = 30/5 = 6.

The rule of thumb: Slow growth = slow moves (longer recovery periods); Rapid growth = rapid moves (shorter recovery periods).

When in doubt, move slowly. Every time you leave a paddock a day early, it cuts a day off the recovery time for all other paddocks.

These facts are easily overlooked. But, if the planning is done on a chart, you will be immediately able to see the loss of recovery times.

Time, Paddocks, and Land Divisions: Overgrazing occurs when animals take foliage before the plant has recovered because they either linger too long or return too soon. Timing and planning on a chart are the keys to avoiding this. Land divisions are the instrument. *The more divisions you have the more recovery time per day of grazing each gets.*

When paddock numbers are high enough that the longest recovery period (slow growth) dictates grazing periods of no more than 3 days, you have the option of using a single recovery period (say 150 days) instead of one for fast growth (say 30 days) and one for slow growth (say 150 days). This simplifies the planning math with *little danger of overgrazing with graze periods of only 3 days.*

Although you would not be likely to overgraze, *you could get better animal performance during rapid growth by dropping to a 1 day grazing period* because the plants and plant parts the animals would be selecting would be less fibrous. This would shorten recovery periods to less than 150 days but that should be ample recovery time for periods of rapid growth.

As paddock numbers increase, grazing periods become too short for overgrazing to occur. *At about 30 paddocks, slow moves during fast growth cease to cause overgrazing. However,* you should remain aware that great damage can occur if animals *do* happen to return to any paddock before the plants are fully recovered. *Remember: Animals moving a day early in a 30 paddock cell will reduce the recovery period in every paddock by a month.* The best way to avoid this is to plan on a chart where you can keep an eye on recovery periods.

If the ranch is under stocked (too few animals to achieve the desired effects), "product conversion" is the weak link. In that case, the best marginal reaction per dollar comes from increasing animal numbers. However, once you reach the point where resource (energy) conversion is the weak link, the return on building more paddocks is amazingly high.

While paddock numbers are low, it pays to find a way to confine animals to smaller areas of land. *Strip grazing within* a larger *paddock* is an example.

At some point, water will become the limiting factor. Like fencing and livestock, water should have money spent on it only when it alone will improve production more than any other measure.

Up to about 30 paddocks, each new division shortens the average grazing period significantly. After that, gains begin to taper off. Any increase in paddock numbers will improve the graze-to-recovery ratio. Also, *any additional paddocks will always increase stock density—the key factor in achieving herd effect.* This is, in turn, beneficial because *partial rest is a greater problem than overgrazing in brittle environments.*

Also, *the rule of maximum density and minimum time applies in any environment.* Even in tropical environments with 100 inches of rainfall, the only way to get reasonable animal performance is by moving animals frequently.

The Effect of Paddock Numbers on Density: Stock density is the number of animals per unit of land (animals per acre)—i.e. Stock density = # of animals ÷ acres in a given paddock. It also represents how many ADAs the herd will take from a paddock every day they are in it.

At very high density, the impact on an acre is very great which means it increases the damage you can do. *Greatly increasing paddock numbers and stock density reduces your risks of drought, flood, disease, etc. but it also increases the penalties you will pay for poor management.*

Advantages of high stock density are:

- Animals graze a greater proportion of available plants more evenly
- Distribution of grazing, dung and urine are more even
- Animals move more frequently to fresh ground which provides a more constant level of nutrition
- Tighter plant communities develop with more leaf and less fiber
- Animal performance improves

Generally, smaller paddocks with higher stock densities will improve faster than larger paddocks with lower stock densities.

If animal performance drops due to either 1) the forage being so rich that it lacks roughage or 2) the rest periods being so long the grasses become excessively fibrous, you need to consider supplemental feeding. Also, abrupt changes in diet can lower performance so watch for this when moving.

Stock Density and Animal Nutrition: Initially, increased stock density may cause a fall off in individual animal performance but overall production and profitability will be enhanced (two 250 lb calves will be worth more than one 500 lb calf any day of the week). However, there are some things you can do to avoid this performance drop.

First is good grazing planning that ensures a more even plane of nutrition. Give the animals the opportunity to balance their diets. The best way to do that is to move quickly to fresh grass—as fast as adequate recovery periods will allow. *Never* force them to eat everything. The only exceptions are when you are using a class of animals whose performance does not matter or when using animals to clean up old material in the non-growing season. You can also consider using supplements to avoid initial drops in individual animal performance.

Forage and Drought Reserves: The traditional practice is to keep some land available as a drought reserve. There are several problems with this. First, forage left standing for long periods of time loses its nutritional value. Second, over-resting grasslands promotes shifts to forbs and woody species. Withdrawing land from production reduces the graze/trample to recovery ratio and puts heavier pressure on remaining land. There is a better way.

Since you can plan how many animal-days a whole cell can supply, *you can incorporate a particular number of days (or months) of grazing for a reserve right into your plan*. To illustrate how this might work, assume you have two, 1,000 acre cells of equal productivity. Both are divided into ten, 100 acre paddocks that are also consistently equal. Herd size is 200 head.

As you can see from the table below, reserving area versus reserving time is equal in terms of forage consumed. But, the result is a quite different treatment of the land. For example:

- Cattle in the acre-reserve cell take 25% more ADA from the land they graze. This means more grass starts recovery from lower on the growth curve—i.e. it re-grows less in the same time than in Cell B
- Given an average 63 day recovery period, grazing periods are two days shorter in Cell B: GP time Cell B = 63/(10-1) = 7 days; GP time Cell A = 63/(8-1) = 9 days

- Cattle move to fresh grass 25% more often in Cell B and therefore perform better

Cell A	Cell B
Graze 800 acres, saves 200	Graze all cells, save time reserve
(200 cows X 365 days)/800 acres = 91.25 ADA	(200 cows X 356 days)/1,000 acres = 73 ADA
Feed reserved = 200 acres X 91.25 ADA = 18,250 ADs	Each acre will have 18.25 AD of feed left (91.25 – 73) = 18.25
	Thus, total reserve is 18.25 X 1,000 acres = 18,250 just as in cell A

There is a simple technique you can use to estimate how many animal-days you have if growth should stop right now. Applied another way, this same technique can tell you how many ADA the cell will have to supply in order to carry your animals for a given time. Then you can evaluate sample areas against that figure and look far down the road as droughts develop as well as plan for the predictable non-growth season.

These calculations involve stocking rate which is the number of acres able to support one animal for the time you expect the herd to remain in the cell. Stocking rate = (acres in the whole cell) ÷ (animals in the whole cell).

Figuring Approximate Stocking Rate: Assume you have a sheep ranch stocked at 1:20 (20 acres/sheep) and winter lasts 180 days. So, 180 AD ÷ 20 acres = 9 sheep days per acre (ADA). That is to say that 1/9 of an acre must feed one sheep for one day. If 1 acre = 4,840 yds^2 then 1/9th acre = 4,840 ÷ 9 = 538 yds^2 and $\sqrt{538}$ = 23. So now you can step off several random squares that are 23 yd on each side and ask, "Will this feed one sheep for one day?"

Now assume that 538 yd^2 is not enough to feed one animal for one day. Work the problem the other way. Pace out sample squares that you believe will support one animal for one day and measure the sides. Let's say that turns out to be 35 yd X 35 yd or 1,225 yd^2 to feed one animal for one day. How many sheep will one acre support for a day? 4,840 yd^2 ÷ 1,225 yd^2 = 3.95 or 4 sheep/acre.

If the ranch is 40,000 acres (with all the roads, steep hillsides, etc subtracted), then you have a total of 160,000 animal days (4 ADA X 40,000 acres) which will feed only 889 sheep (160,000 ADs ÷ 180 days = 889).

So, if you are currently running 2,000 sheep, you will run out of forage in about 80 days (160,000 ÷ 2000 = 80).

Factoring in Physiological State and Wildlife Needs: If it makes you feel better, you can use tables to convert standard animals units to allow for the physiological state of the animals. But, in practice, it is just as effective to convert SAU based on weight or size only and factor in physiological state when assessing forage samples in the field. Do this by altering the question:

- *Could this square feed one animal today?* If all your animals are dry pregnant cows and you have no wildlife.
- *Could this square feed one animal **comfortably** today?* If your cows are lactating or you have steers that need to gain weight.
- *Could this square feed one animal comfortably today and leave enough forage for the wildlife?* If you also have a significant wildlife population.

- *Will this square feed one animal and leave plenty of cover?* If you want to leave cover for game birds.
- *Will this feed one animal today and leave plenty for litter cover?* If you are trying to cover the soil as quickly as you can.

When asking these questions, think like whatever animal you are judging for. Imagine you have a sack the size of the animal's stomach and you have to fill it in a limited time with fresh material and you can only use one hand to do it. Roughly, imagine yourself filling a sack with a half a bale of hay while picking with only one hand. If that seems it would be difficult, then it would probably be difficult for the animal. If you are running mixed herds in different stages of lactation or breeding, use common sense.

When a cell supports livestock only part of the year, you have only the animal days required by wildlife to worry about the rest of the time. This takes experience which you get from old grazing charts, monitoring and fixed point photos.

For animals that browse and graze, look at all the vegetation. Also be aware that animals will take no old oxidizing material and have a limited time for feeding. Consider the quality of the forage as well as the bulk. With practice, you will become increasingly accurate. If you find yourself fudging or in doubt, call it a "no."

After sampling several squares and finding that, on average, they would not provide the required feed, the next step is to calculate how many animals you can run. Do this by two or more men pacing off a sample square they believe would feed on animal for one day. Although this is crude, it is likely the best information you can get. *Get multiple opinions and always err on the conservative side.*

Taking all the non-growing season ADs planned for one paddock in one grazing at the beginning of the season would provide you a precise reading on your estimate and allow you to make appropriate adjustments in other paddocks.

Determining Correct Stocking Rates: The technique described above can also help *during the growing season*. Start by *sampling the longest recovered paddock that animals are soon due to graze. Compute the size of the square necessary to feed one animal for one day during the shortest grazing period allowable for that paddock*. Since you would only use that size square under *ideal* growing conditions, you are surely overstocked if it does not pass the test. *For example*, suppose you plan for 800 cows in a 600 acre paddock on 3 to 9 day grazing periods: (800 cows X 3 days) ÷ 600 acres = 4 ADA; 4,840 yd^2 per acre ÷ 4 ADA = 1,210 yd^2 per animal day; $\sqrt{1,210}$ = 35 yd per side of an animal-day square.

If this would feed one animal for a day, compute the *smaller* square that would support one animal for one day given the longest grazing period in that paddock. This is 9 days in our example. So, (800 cows X 9 days) ÷ 600 acres = 12 ADA; 4,840 yd^2 per acre ÷ 12 ADA = 403 yd^2 per animal day; $\sqrt{403}$ = 20 yd per side of an animal-day square. *If that doesn't pass, you are gambling on rapid growth. Monitor diligently.*

If animals deplete forage or start to eat litter before they are scheduled to move, suspect overstocking but check other factors:

- Did you overestimate the paddock's capacity? Taking ADAs from a better paddock might relieve the pressure.
- Low stock density can produce forage too rank to eat. Mowing, burning or herd effect can solve that problem but plan for higher density as soon as possible.

There is a way to train to judge more accurately. Fence off some sample squares in one of the paddocks the size that you estimated it would take to feed one animal for one day. Then subject the paddock to that average grazing pressure estimated and compare the grazed land with the enclosures.

The Critical Non- or Slow-Growing Season: The amount and quality of standing forage at the start of the non-growing season is what determines your stocking rate. No other factor tells you as much about the carrying capacity of the land.

If you do not carry a herd during the non-growing season (yearling operation, for example), regulate your numbers to leave enough forage: 1) for wildlife and 2) to provide a reserve against a late return of growing conditions.

If you run animals year-round, you will have to allow for enough to carry you through the non-growing period *and* provide for wildlife and drought.

A dry summer typically does not leave enough forage to last through the winter. But, good planning can compensate by treating every year as though it is going to be a dry one.

Furthermore, droughts are mostly a result of non-effective rainfall rather than insufficient rainfall. The higher the percentage of bare soil, the less effective is the rainfall and the more likely you are to have a drought.

If You Run Out of Feed…or Even if you think you might be:

- Before cutting stock, do field checks and determine how many animal-days you actually have.
- Consider changing your plans. This usually means combining stock into fewer, larger herds. Consider combining all herds into one to extend recovery periods and keep animals moving as frequently as possible.
- Consider various ways to use forage more efficiently—e.g. increasing paddock numbers with temporary fence, using portable water, using supplements, etc.
- If you must destock, the sooner you destock, the less you will have to destock.

The Economics of Destocking: If you must destock, the sooner the better. You can get rid of 1 cow today or you will have to get rid of 180 cows in 180 days from now because ever cow will use one cow-days worth of forage each and every day.

Nutrition during Non- and Slow-Growth Periods: Say you have enough forage after the first freeze to last you till spring. But, nutritional value drops steadily under grazing. You can buy supplements for the livestock but the wild animals will go short. Under continuous grazing, supplementation will be necessary after half the forage has been selected and the need will continue to increase until new growth sets in. But there is a better way.

The more paddocks you have the better you can intelligently plan forage use through the non-growing season and drought reserve period. For example, you can plan your non-growing season grazing so that *when animals enter a new paddock, they also have access to previously grazed paddocks. This enables them to better balance their diets.* (Since the plants are dormant, you don't face an overgrazing problem by doing this.)

Suppose you have 100 paddocks and the herd stays only two days in each—i.e. they move to fresh ground every other day. Forage use is efficient and the supplement needs are minimum.

If the drought is even worse than you planned and you are forced to make another pass through the paddocks, there is no doubt but what you face a depleted nutritional selection. But, you will always be better off than if you had planned for the same situation by reserving un-grazed areas.

If a drought does not occur, the accumulated grass is usually not a problem in low rainfall environments. In fact, it can actually be very nutritious from growing on highly mineralized soils.

Managing a Drought: Always assume the growing season will be a poor one—either because *spring rains do not materialize* or because *they are all you get. A time reserve will get you through the former but, in case of the latter, you will have to re-plan all the way through to the start of the next growing season.*

Never wait until the end of the growing season to take action. *At the first indication of a summer without rain, estimate animal-days to get an idea of how long you can last.* Review all the measures that might help get you through like *combining your herds* and moving them through all the paddocks you have.

Watering Large Herds: Lack of water is the most common reason given for failing to increase herd size. It is important to plan for adequate water when you create your land plan. Some solutions include larger storage reservoirs or piping water to areas where it will be needed.

If you need to amalgamate herds (during a drought, for example), you can *provide temporary water by hauling it or driving animals to water every second or third day.* Even breeds that did not originate in arid areas can adapt to going a long time between drinks. They must be used to a routing and *never* disappointed.

Creating Herd Effect: The tool of choice for offsetting partial rest is animal impact. Animal impact *comes from either stock density or herd effect. Stock density is the number of animals per unit area of land. Herd effect is a matter of herd size and animal behavior.* In brittle environments, stock density alone cannot overcome partial rest unless pushed to extremes. Herd effect is what is needed and *the bigger the herd the better.*

In grazing planning, anticipate the areas where you will need herd effect to:

- Suppress brush by breaking it down
- Return un-grazed plant material to the soil
- Promote tighter spacing between plants, holding more litter in place and thus causing more water to soak in.
- Soften the banks of gullies and start plant growth
- Reduce noxious weeds by direct impact and creating conditions favorable to grasses
- Clear firebreaks or roadsides

In the past, pack hunting predators ensured herd effect. But, animals spread and remain calm when free of that danger. Driving livestock with cracking whips or dogs causes herd effect but costs in terms of lowered animal performance. One Wyoming rancher used dude ranch guests on horseback to bunch cattle and move them to fresh pasture each summer day.

Attractants are the most practical way.

- Supplements fed on the ground

- Salt that a herd has been denied for some time. Granulated and fed on the ground.
- Diluted molasses sprayed on weeds or firebreak areas.
- Static inducements such as salt blocks and minerals do not produce herd effect. Putting them on a trailer that can be moved works better but is still short of the ideal.

The downside is the small area that is usually affected; however, attractants are the most practical for healing eroding gullies, thinning dense brush thickets and other site-specific objectives.

Training has a role. Animals who have never tasted molasses will not recognize its smell. Livestock will learn to come to a whistle. This not only stimulates herd effect but also simplifies moves to new paddocks. Including a few trained animals with untrained stock speeds the training. Calmer stock handling methods (as developed by Bud Williams) allow concentration of animals without stress.

Single vs. Multiple Herds

Each additional herd carries hidden costs. If you must pull bulls and heifers out of the herd, run them on continuous graze in a sacrificial paddock that changes from year to year. This leaves the bulk of the paddocks available for the main herd.

Single Herd Management: The greatest obstacle to combining herds is worry about handling so many animals and handling classes or species together.

Specialized breeding programs may require separate herds but large numbers of animals do not. Animals can learn behavior that makes herd size irrelevant to handling. Also, the layout of facilities and calm handling are more important than herd size.

Successful cattle herds can include all age and sex classes of cattle—i.e. bulls remain in the herd year round and calves are weaned without being separated from their mothers. Once females are pregnant, why remove the bulls? If you want the caving period to last only 60 days or do not want young heifers to breed too early, regulate that by culling.

You can even handle single-sire breeding without many herds. A 500 head registered Simmental operation ran a single herd. Each bull served the cows only once and the rancher could run up to 90 cows per bull. The cost saving allowed him to purchase better bulls. This was accomplished with a radial cell with specifically designed facilities in the center. (That is described under the lesson dealing with infrastructure planning. This proved to be very profitable despite the fact that he had to feed the bulls over a 2 month period. Fewer and better bulls more than paid for the feed.

Production per animal is important. Production per acre is more important if you are serious about profit.

Use your creativity.

Multiple Herd Management: There may be times when running two or more herds make sense.

Deciding the Number of Herds (an example): Assume you have 600 steers on 3,000 acres divided into 56 paddocks of 54 acres each. You would like to push 100 steers ahead and market them early. You have four possible strategies:

1. Keep all 600 together and select the best 100 at sale time

2. Split them into two herds and push them through all the paddocks separated by the pertinent recovery period. This allows the 100 head greater ability to select an ideal diet.
3. Split them into two herds in two cells with the 100 being run through the best 20 paddocks and the 500 using the other 36.
4. Let the 100 make the first selection in each paddock with the 500 following right behind with no break between the two herds.

Consider:

- Land benefits from larger herds, high density and short grazing periods
- Livestock benefit from rapid moves to fresh grass and lower ADA take.

Think first about the land and then about the livestock:

- Strategy 1 is best for the land—lowest grazing pressure (ADA harvested per grazing period), highest stock density and shortest grazing period. Strategy 4 is second best.
- From the cattle's view: Strategy 4 is best for the 100 but not the main herd which would be slightly better off with Strategy 1.
- 2 and 3 are poor for the land—forage production and rainfall effectiveness would decrease.
- For the cattle, #2 is likely better than 3 for both herds.

Wild Grazers and Browsers: Plan to avoid areas where game are rutting or fawning and only lightly graze areas where cover is needed for ground nesting birds. Some wildlife may graze the same paddocks as livestock do in the growing season. In fact, they may actually link up with the domestic stock because forage one or two moves behind the livestock is re-sprouting and very nutritious.

The best that you can do is to make every effort to maximize animal impact while minimizing stock time in paddocks. Expect wild animals to associate with the herd in the growing season by grazing behind it on re-growth. Monitor carefully.

Matching Animal Cycles to Land Cycles: The argument over the best time to calve and lamb is never ending and tradition often outweighs all the other factors. Tradition dictates that calves are born in early spring to be marketed in the fall. Why not summer calves?

The standard worksheet allows you to compare alternatives. Complex decisions are best sorted out with gross profit analysis that considers each breeding/production policy as a separate enterprise.

Keep in mind that cows need a rising plane of nutrition for conception and above average feed for six months of lactation. Rethinking the cycle of forage may even result in a realization that keeping stock through the winter doesn't pay.

When you actually do your planning, no tradition need be sacred. Further, as your land improves, new possibilities will open—i.e. more cool-season grasses, extended grazing seasons due to improved water cycles, fibrous plants giving way to more palatable species, etc.

All decisions should be tested toward your strategic goal.

Pests, Parasites, and Other Headaches: You can manipulate any situation characterized by a strict routine by changing your own routine, if you can find the critical point. Here are some common examples:

- Parasites (like liver flukes) leave their host for part of their cycle. Records will show were the herd was at that time. Plan to have your animals elsewhere at that time.
- Some pests breed in fresh manure. Moving animals frequently onto fresh ground mitigates this.
- Predation increases when predators have good cover, insufficient wild prey, and young to feed. Plan to calve on safe sites when natural prey is plentiful and predator needs low.
- Predation is higher when livestock are spread thin over large areas. This can be reduced by running larger herds and/or mixed herds with dogs or llamas.
- When floods occur in predictable seasons, plan to graze flood prone areas afterward. The un-grazed plants will slow the flow and catch sediment.
- Poison plants are a threat for limited periods or when other forage is scarce. Graze these areas at safe times or with very light grazing in susceptible times.
- Ground-nesting birds need certain habitat during nesting or breeding. Keep stock out of those areas during those times.
- If grass fires threaten certain areas, graze them early.

Practical Application: Creating the Grazing Chart

Introduction

Planned grazing is done according to a checklist but one which differs from a simple checklist in that it gives a sequence for making decisions that takes into account the effect of one decision building on another.

Guidelines for planning

- Follow the steps on the checklist in order.
- Plan on paper focusing on the recovery periods (rather than grazing periods) which can only be seen on a chart.
- Keep your first plans simple—minimize the number of herds and avoid sophisticated grazing strategies until you master the basics.
- Plan pessimistically on any point on which you have any doubt.
- Plan creatively every time. Easy planning year after year will tempt you to abandon the process.
- Create one plan per herd for the growing season and one plan per herd for the non-growing season and drought reserve.
- Aim to maximize forage production during the growing season.
- In the slow or non-growing season, aim to ration the forage grown and return old standing forage to the ground.
- Monitor the plan. No plan ever goes exactly.

Checklist for Planned Grazing

As stated above, you should:

- Plan, monitor continuously, control deviation as soon as possible and re-plan when necessary. *Livestock operations usually call for major planning twice a year.*
- Make your *first plan at least a month before the main growing season begins.* You will be trying to grow as much forage as possible and *you do not have to plan to a specific date. The plan remains open because you don't know when growth will end.*
- *Make the second plan toward the end of the growing season when forage reserves become known. Ration the forage over the months ahead to* a theoretical end point *a month or more after your most*

pessimistic estimate of when new growth could occur. This becomes your drought reserve. Plan to a specific date—i.e. this is a closed plan.
- In areas *where rainfall is low and unreliable, overlap the drought reserve in the closed plan with the potential start of the next growing season.*

THE OPEN ENDED PLAN

As you progress through the steps, constantly refer to the following: Blank Grazing Plan from HMI, Sample Grazng Chart (Open-Ended Plan) 2 pages, and Chart With Grazings Plotted (Open-Ended Plan) 2 pages. Out of the three, you should be able to locate what you are reading about.

Sample Grazing Chart (Open-Ended Plan)

Grazing Plan (Livestock/Wildlife)

YEAR: 2005   PLAN: (OPEN-ENDED) CLOSED

Columns 1 ADA/H Actual/Estimate	2 Estimated Relative Paddock Quality	3 Paddocks Size	Number/Name	MARCH	APRIL	MAY	JUNE
/		2000	A				
/		850	B1				
/		870	B2				
/		2200	C				
/		560	D1				
/		250	D2				
/		370	D3				
/		320	D4				
/		2500	E1				
/		400	E2				
/		750	F1				
/		1000	F2				
/		200	H1				
/		450	H2				
/		400	H3				
/							
/							
/							
/							

Calving spans March–May.

				MARCH	APRIL	MAY	JUNE
		21. Rainfall					
		22. Snow					
23. Growth Rate (F/S/O)							
24. Supplement or Feed—Type and Amount							
25. Number/Size of Herds				1/750	1/1030	1/1030	1/944
26. Paddocks Available				12	12	12	12
27. Recovery Period(s) or Number Selections				30 - 90	30 - 90	30 - 90	30 - 90
28. Avg GP or AMGP/AMxGP Type of Animals				No. / Average Weight / % Unit / Total Units	No. / Average Weight / % Unit / Total Units	No. / Average Weight / % Unit / Total Units	No. / Average Weight / % Unit
29.							
30.							
31.							
32.							
33.							
34. Total SAUs							
35. Cell Size: 13,120				Remarks:	Continuous grazing paddock		Loco weed
36. Stocking Rate: 1:13					Crops		Severe erosion requires anim
37. Avg Annual Precipitation:							
38. Season Total Precipitation:							

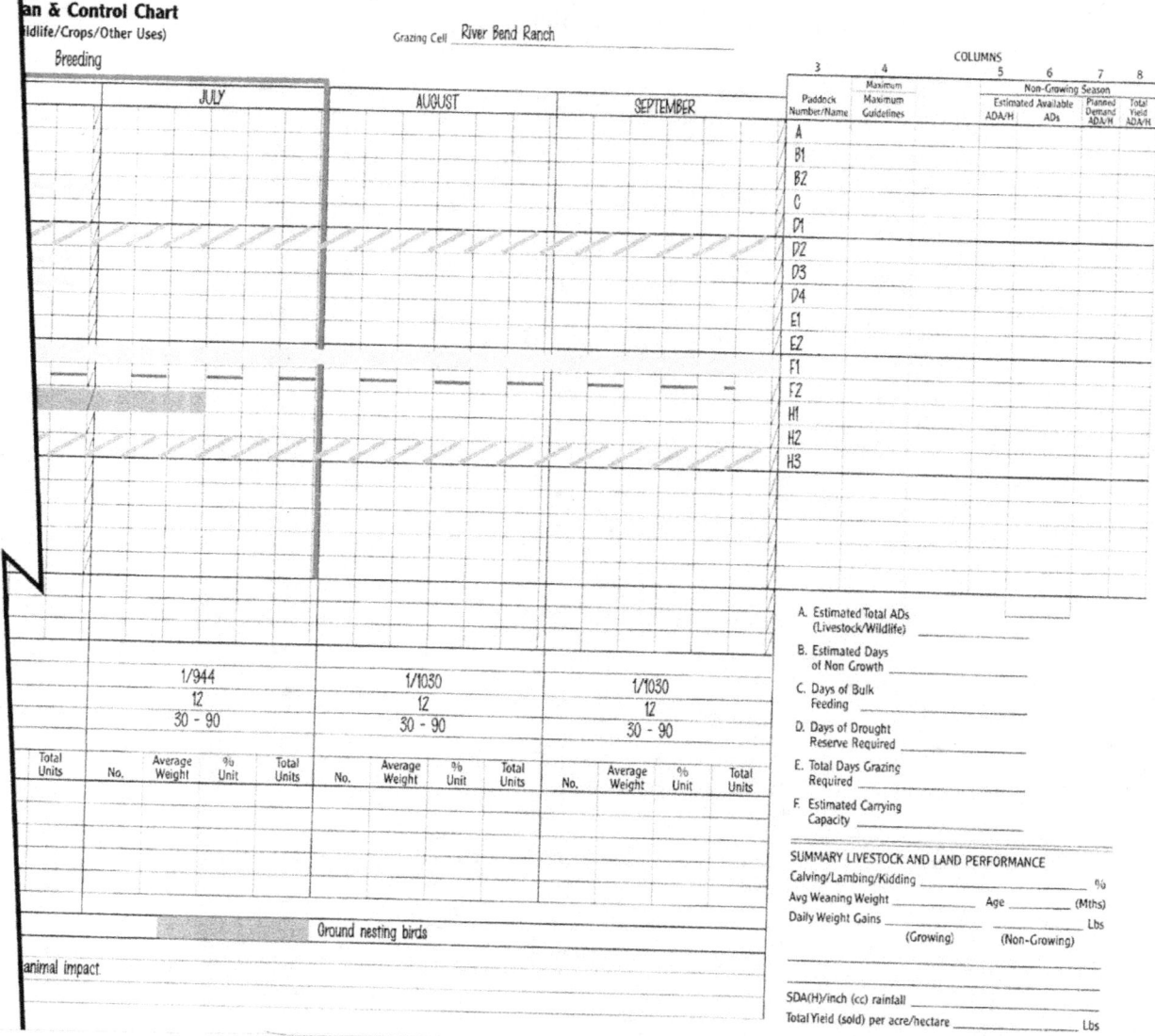

Chart With Grazings Plotted (Open-Ended Plan)

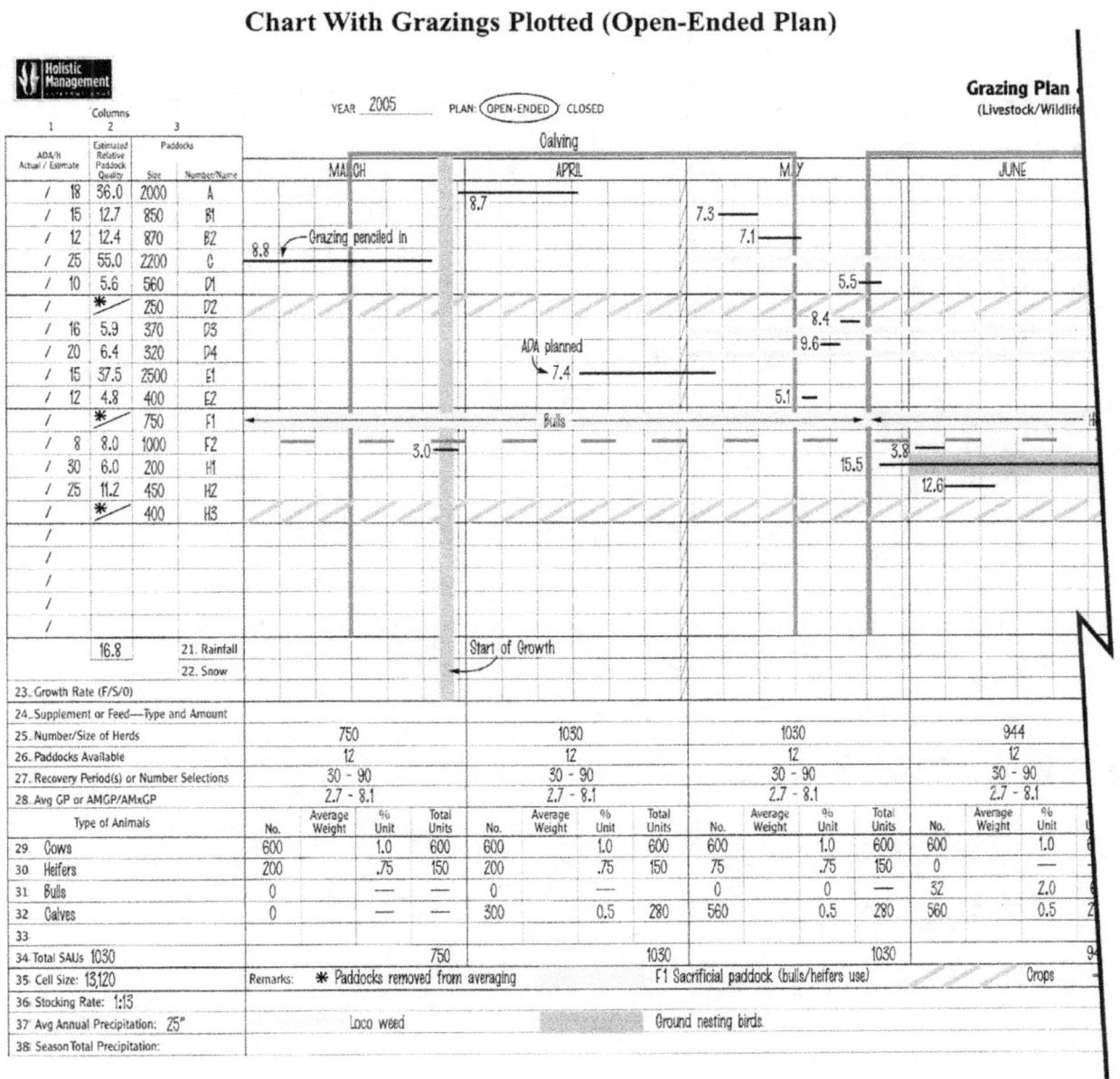

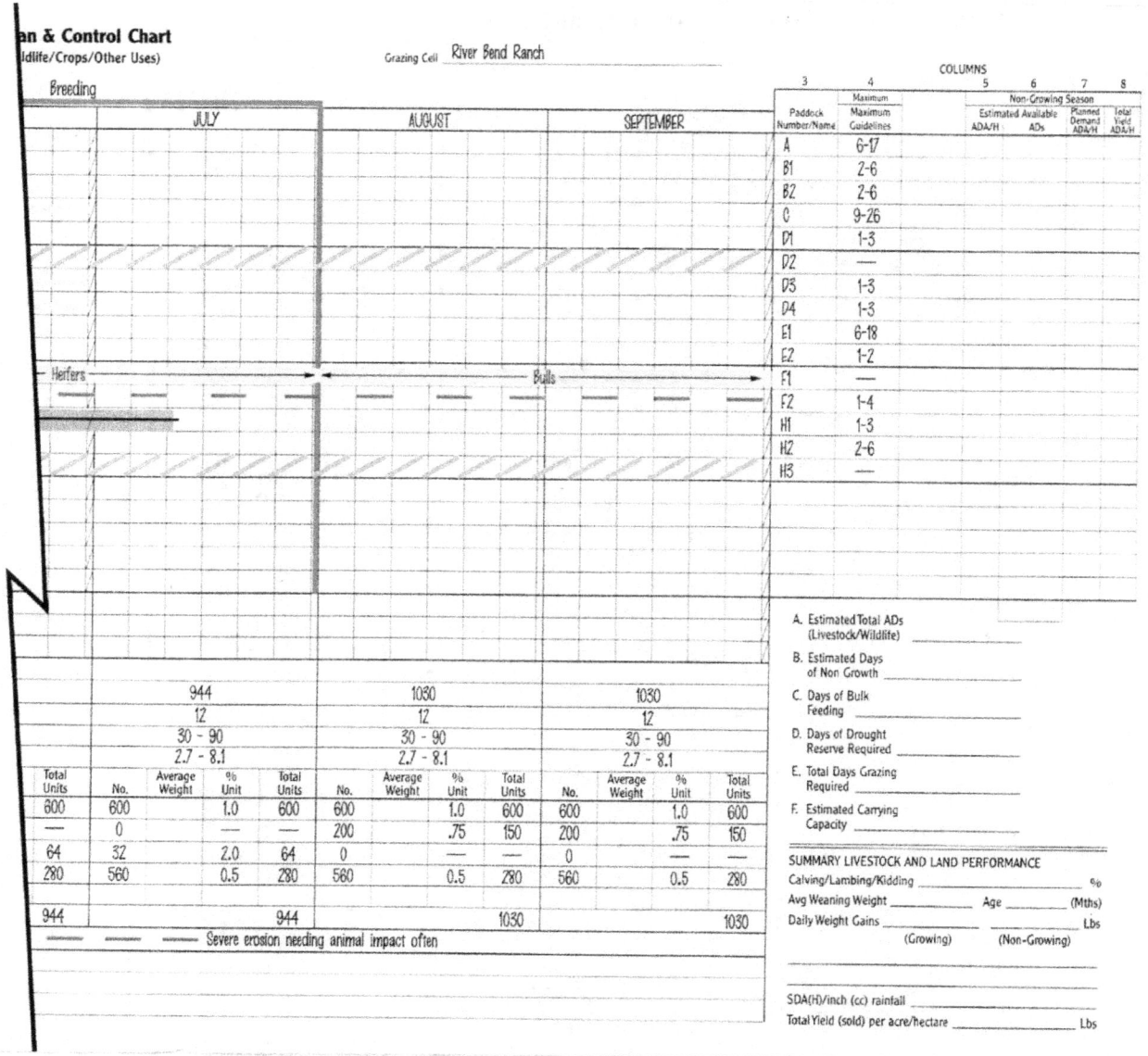

Step 1. Make Initial Decisions: Look at the Big Picture

Think of all the factors that may have an influence. Remember, in the growing season your aim is to maximize forage production and in the non-growing season it is to ration the forage on hand and trample old, standing forage onto the soil surface. Write down all the things you can think of that need to be taken into account.

Next, think over the months you are planning for and envision the whole ranch. *Answer the following questions on paper:*

- Is the entire area one grazing cell with one herd or different cells for separate herds? The answer determines how many plans you will need. The fewer the herds, the better.
- What stocking rate do you plan to use? This could change in the course of planning.
- What crops will be planted and where? Use a worksheet to schedule planting and keep it handy.
- What drought reserve will you hold and when would you start to use it based on the area's history?

- Has your biological monitoring identified any problems? Do you need increased animal impact in some areas, rest on others, or reduced partial rest all over?

Step 2. Set Up the Grazing Chart

Use one planning chart per cell/herd. If it is a breeding herd and you cannot be convinced that herd sires or young females can be left in the herd year-round, consider allocating a sacrificial paddock that will be different every year. Bulls would spend all but the breeding period continuously gazing that paddock. Heifers could be moved to it over the breeding period. This essentially amounts to one chart since you have one herd moving through all paddocks but one.

- If you have more than twenty paddocks or are planning for more than seven months, cut and paste enough sections from other charts.
- Record the year and name of the cell and circle "open" or "closed" ended at the top.
- The blank row at the top is for the months.
- Enter paddock numbers in both column "3"s and their sizes in the left column 3.
- If crop fields will have stock, transfer those to the chart as paddocks.
- Avoid marking in anything not described in the steps—i.e. keep the chart uncluttered.

Step 3. Record Management Concerns Affecting the Whole Cell.

Indicate all the management events in color code. Draw vertical lines through all the paddocks on the starting and ending dates and connect them across the top of the chart. Write the event on the top connecting horizontal line:

- Livestock events: breeding, calving, weaning, etc.
- Use the list of planning factors you created earlier—wildlife, hunting seasons, etc.
- Enter the times when critical people will be away.

Step 4. Record Herd Information

- Enter the types of animals and their numbers each month in rows 29 to 33. If you are running multiple species and/or several classes of stock, record two to a line and separate the numbers with a slash. An option would be to cut and paste additional rows.
- Convert stock numbers to standard animal units (SAUs). Record the total for each month in row 34. Record the *peak* figure in the first column (this is the *maximum* SAUs entered on row 34). Do not include animals in a sacrificial paddock in the SAUs except in the months they will be joining the main herd.
- Enter total cell size in row 35 (this is the sum of the paddock sizes in column 3). Calculate the stocking rate by dividing cell size by the peak SAU and record it in row 36.

Step 5. Record Livestock Exclusion Periods

Draw a color coded horizontal line through the time periods when a paddock cannot have the herd in it due to lack of water, annual flooding, hay cutting, etc. If you plan to remove herd sires to run in one sacrificial paddock, note which one. Use a different color for each factor and explain the meaning under "Remarks" at the bottom. (Bulls continuously graze paddock F1 - Yellow; Severe erosion in paddock F2 requires animal impact--broken red; Ground nesting birds in paddock H1 -- blue, etc).

Step 6. Open-Ended Plan Only: Check for Unfavorable Grazing Patterns

- If this is your first year, move on to the next step. Otherwise, look over past grazing charts and mark paddocks repeatedly received an "H" (for heavily grazed) in early or late past growing seasons. To avoid repeating the error, exclude the paddocks for the early and late periods this year.
- Check for paddocks that failed to receive adequate recovery time in the last growing season and avoid those early in this growing season.
- If the growing season has begun early, leaving you with "drought reserve," remove the old growth before it affects new growth or livestock performance.

Step 7. Record Paddocks Still Available

- In row 26 under each month note the number of paddocks still available. If a paddock is available for more than 50% of the month, count it for the whole month if this is your first time to plan. You can graze the paddock on either side of the unavailable period which will keep paddock numbers higher.
- Once you have experience with recovery periods, you will likely count a paddock as available if the unavailable period is less than the average recovery period. For example, if a paddock is marked unavailable for 50 days and recovery ranges from thirty to ninety days (average of 60 days) count it as available.

Step 8. Note Paddocks Requiring Special Attention

Mark through any paddock that needs special treatment with a color-coded highlighter and explain the colors under "Remarks" at the bottom. Factors might include:

- Sacrificial paddock subject to continuous grazing;
- Bare ground or noxious weeds requiring herd effect:
- Areas needing more rest to create brush for wildlife;
- Areas to remove forage toward the end of the growing season for fire prevention;
- Paddocks you plan to strip-graze.

Think about limitations—proximity to crops or neighbors bulls, poisonous plants, lack of shelter, parasite cycles, etc. Review the list of planning factors you created earlier. Note wildlife management factors, multiple-use factors like logging operations or hiking trails on public lands.

Step 9. Rate Paddock Productivity

Rate Forage Quality: For your first *open-ended* plan, use a 1-10 scale *quality* rating for each paddock with the best one being a 10 and all the others relative to it. Record your rating to the right of the slash in column 1.

This provides a good approximation but you should switch to either of the following methods as soon as possible

- *In each paddock, pace off several random squares that could feed one animal for a day and compute the area in square yards. Be pessimistic. Average the answers to get an estimate of ADAs. This is difficult if the terrain is varied and you are new. One way to allow for this is to remove all the acreage covered by the inaccessible areas. Also, you can sample within the poorer grassland areas more than the good ones.*

- *Open ended plan*: Transfer these figures to the right of the slash in column 1

Factor Forage Quality and Area into Paddock Productivity: If any paddock is unavailable for the entire period (sacrificial paddock) do not include it in the ratings calculations.

Open-ended plan

1. Multiply the paddock rating (col 1) by the size (col 3). Divide by 1,000 (to convert the number to thousands for simplicity) and enter the result in column 2.
2. Now add the figures in column 2 and divide by the total number of *available* paddocks. Enter this in the box at the bottom of column 2. (See 16.8 in the box beneath column 2 on the Open-Ended Plan)

Step 10. Open-Ended Plan Only: Determine the Length of Recovery Periods

- Determine a range of recovery periods to allow for fast and slow growth. Recovery periods are the times that severely bitten plants need to recover depending how fast they are growing. *The faster the growth, the shorter the recovery period needed. In arid areas 30 to 90 days are usually enough. In areas of higher precipitation, 20, 40, or 60 may do. Irrigated pastures might take only 15 to 30. These are only guidelines. Prolonged adverse growth conditions may lengthen the time to 150 days or more.*
- Record the minimum and maximum recovery periods on row 27 under each planned month.
- *If you have a great many paddocks (enough to ensure grazing periods of 3 days or less)*: you can use a single recovery period. A recovery period as long as (or longer than) the entire growing season will allow plants to recover regardless of growth rates. If your growing season is 180 days and you have 100 paddocks, grazing periods would average 1.8 days.
- *If you have very few paddocks*: Grazing periods become so long that plants will be overgrazed no matter what you do. But, overgrazing during the grazing period is less harmful than overgrazing after an inadequate rest period. So, plan to achieve your longest rest period (i.e. 90 days if that is what it is) with one grazing. Meantime, try to get paddock numbers up to at least 10 as soon as your finances and weak link analysis will allow.

Step 11. Open-Ended Plan Only: Calculate Grazing Periods

- Calculate *average* min and max grazing periods for the cell and record in row 28 by dividing the respective recovery period by the number of paddocks *available* (line 26) minus 1.

$$\text{Average minimum grazing period} = \frac{\text{Minimum recovery period}}{\text{Number of paddocks -1}}$$

Example: 30 (min recovery period) ÷ (12 – 1) = 2.7 (See line 28 of the Open Ended Plan)

$$\text{Average maximum grazing period} = \frac{\text{Maximum recovery period}}{\text{Number of paddocks -1}}$$

Example: 90 (max recovery period) ÷ (12 – 1) = 8.1 (See line 28 of the Open Ended Plan)

- Convert these to *actual* min and max for each paddock according to paddock ratings in col. 2.

$$\text{Min (max) grazing period} = \frac{\text{Paddock rating X average min (max) grazing period}}{\text{Average paddock rating}}$$

Example: Minimum: 37.5 (paddock E1 rating) X 2.7 (average min grazing period) = 101.25 ÷ 16.8 (average paddock rating) = 6.03 (rounds off to 6). Maximum: 37.5 (paddock E1 rating) X 8.1 (average max grazing period) = 303.75 ÷ 16.8 = 18.

This gives you the actual grazing periods to be used when growth is fast (min) and when it is slow (max).

- Record these rounded off into whole numbers in column 4.

- If some paddocks have much longer grazing periods than others, check to see that recovery periods are adequate.

    o Add all the *minimum* grazing periods together
    o From this total, subtract the longest minimum grazing period to find the *actual* recovery periods for these paddocks
    o If any recovery period is much too short, add days to the minimum grazing periods in other paddocks that can absorb them.
    o Do the same for maximum grazing periods but the problems are less critical if you can't make complete adjustments.

This figure is a segment of the planning chart showing columns 2, 3 and 4 of the Open Ended Plan. It illustrates how minimum and maximum recovery periods are calculated, recorded and adjusted if need be.

Figuring Grazing Periods (Open-Ended Plan)

2 Estimated Relative Paddock Quality*	3 Paddocks SIZE	3 Paddocks NUMBER/NAME	4 Minimum Maximum Guidelines
36	2,000	A	6 – 17
12.8	850	B1	2 – 6
12.4	870	B2	2 – 6
55	2,200	C	9 – 26
5.6	560	D1	1 – 3
3.8	250	D2	—
5.9	370	D3	1 – 3
6.4	320	D4	1 – 3
37.5	2,500	E1	6 – 18 Days
4.8	400	E2	1 – 2
8.3	750	F1	—
8	1,000	F2	1 – 4
6	200	H1	1 – 3
11.3	450	H2	2 – 6
5.2	400	H3	—
16.8		21. Rainfall	

*ADs in thousands; bottom figure (16.8) is average ADs

28 AMGP / AMxGP	2.7 – 8.1	2.7 – 8.1

Minimum recovery periods in the highest rated paddocks will be

Paddock	Recovery
A	33 – 6 = 28 Days
C	33 – 9 = 24 Days
E$_1$	33 – 6 = 27 Days

If the 6 days lacking in paddock C are distributed among any of the paddocks that have sufficient recovery time, all paddocks will have 30 days or more.

$$\frac{37.5}{16.8} \times 2.7 = 6.0$$
$$\phantom{\frac{37.5}{16.8}} \times 8.1 = 18.0$$

Total of min. GPs = 33

- If paddock numbers change during some months (see row 26), you will have two sets of numbers. Record the second set in the blank column to the right of column 4. Be sure your average paddock rating is based on the *paddocks actually* used and *not* on *the total number* of paddocks.
- If numbers change several times, base grazing period calculations on the average number of paddocks. You will make adjustments to these as you plot them in Step 14.
- *If you have one long recovery period (lots of paddocks),* the average grazing period will be only a few days and the mathematics are essentially the same:

$$\text{Average grazing period} = \frac{\text{Recovery Period}}{\text{Number of Paddocks} - 1}$$

- Record the rounded off to one decimal place figure in row 28.

- If you have only a few paddocks but strip graze within each of them, the figure will show many days of grazing due to the long recovery period. However, each strip is grazed for only a few hours up to a day. Convert average to actual grazing period by factoring in paddock quality:

$$\text{Grazing period} = \frac{\text{Paddock rating X Average grazing period}}{\text{Average paddock rating}}$$

Step 12. Not Applicable - Closed Plan Only

Step 13: Not Applicable - Closed Plan Only

Step 14. Plot the Grazings

Mark the moves of the herd on the chart showing the length of the grazing period by the length of the penciled line. [See: Chart with Grazings Plotted (Open-Ended Plan)]

Other guidelines:

- Color-code events to be sure animals are at the right place at the right time.
- Indicate events (such as calving) to prevent young from being separated from their mothers. For example, with gates along the fence lines, you could plan to move animals trough adjacent paddocks. Or, with gates only at the cell center, you might move animals through opposing paddocks. In either case, leave gates open between moves so mothers can go back to hidden young.
- *Plan grazings backward from periods when livestock nutrition is critical.* Reserve paddocks with high quality forage and plan which paddocks animals should come from to get there.
- Color-code problems or special management concerns—poison plant danger, need for cover for ground nesting birds, etc. Might want to drop the grazing period to the minimum or even lower.
- If you drop to the minimum in several paddocks, check recovery periods. You may have to lengthen grazing periods to ensure recovery times are adequate.

For the Open-Ended Plan:

- Use the longest (maximum) grazing period in column 4 so as to make the plan conservative.

- Constantly watch recovery periods as you plot. Every day taken off of one paddock's grazing period takes a day off of *every* paddocks recovery period. *The greatest danger of overgrazing is from recovery periods that are too short...not grazing periods that are too long.* When in doubt, slow down.
- If paddock numbers change over the season, make sure to adjust the grazing periods to ensure adequate recovery times.
- Plan the grazings well into the growing season with open-ended plans. If the season turns out well, keep extending the plan. If it turns out to be a dry one, close your open-ended plan and immediately create a closed plan.

Step 15. Make a Final Check of Your Plan

How to factor in the physiological state of the animals or changes in herd size: Calculate the ADA you plan to take out of a paddock during each grazing by looking below the penciled grazing line to row 34. Multiply the SAUs by the number of days in the paddock. Divide the total by the size of the paddock (Col. 3). Write this figure lightly in pencil just to the *left* of the grazing line. For example: Consider paddock A on the Plotted Chart: You plan to graze this 2,000 acre paddock in April. During that month you will have 1030 SAUs (line 34) and you plan to spend the maximum of 17 days in the paddock (from Column 4 calculated in the figure above). So, 1030 SAUs X 17 days = 17,510 SAU-Days divided by 2000 acres = 8.7 ADAs

Open-Ended Plan: Use common sense. Adjust the grazing periods based on paddock quality and the physiological state of the animals. If grazing pressure is too high in a paddock rated as poor, reduce it by a day or so. If lactating cows are being bred, move them more quickly. Add about the same number of days to the other paddocks that can take more grazing. You are attempting to even out the plane of nutrition but are only guessing. So…

Make a field check of sample areas to see if they can feed one animal for one day. Get the sample area size by dividing 4,840 yds^2/acre by the ADA to the left of the grazing line and then take the square root of the result. This gives you the length (in yd) of each side of the square needed. Factor in the animals' physiological state by qualifying the question: "Would this area feed on cow *comfortably* today?" If lactating cows are to be bred on a rising plane of nutrition, ask "Would this area feed one cow comfortably *and with forage to spare?*"

Continuing with the Paddock A example: Divide 4,840 yds^2/acre by 8.7 ADAs and you should get 556.2 square yards. Taking the square root, you get 23.59 yards. So pace off a square that is about 24 yards on each side and apply the appropriate test as explained just above.

The plan is complete!

Step 16. Implement (and Monitor) the Plan

You have produced the best plan possible. However, circumstances change over time and adjustments will have to be made.

Open-Ended Plan Guidelines:

- *Monitor daily growth*. Drop to the minimum grazing period when growth becomes rapid (unless your grazing periods are already short enough—one to three days). Compare grazed plants in the paddock you just left to the paddock you are now in to help judge the rate of re-growth. Also, mark

grazed plants with flagged wire stakes or place portable wire cages over some plants to later compare with the grazed ones surrounding them.
- ***Growth rates are seldom rapid for more than a few days at a time.*** As growth slows, move back toward the longer grazing period (column 4). If in doubt, assume slow. *Color of the grass also indicates rapid growth.* If growth is rapid for such a prolonged time that grazings shift far off the plotted moves, re-plan from that day forward. This normally only involves changing the plotted moves (which is why you should plot the grazings in pencil).
- *When you have enough (many) paddocks for a long recovery period, you do not have to monitor growth rates.* But you still need to monitor livestock, forage bulk, wildlife concerns, etc.
- With low paddock numbers, overgrazing happens either when stock stay too long during fast growth or return too soon during slow growth.
- *With higher paddock numbers, the greatest danger is returning too soon during slow growth.* This rarely happens when you have 100 or more paddocks but you still need to keep an eye on recovery periods.
- Move livestock as slowly as nutritional needs and the maximum grazing period permits when growth rates are slow
- Animals perform better when they move faster but if they move too fast, overgrazing will result.
- If you can look ahead and see that animals are going to return to paddocks before plants have recovered, slow down.
- If you find that animals run out of forage in any paddock, you have misjudged the paddock.
- If you run out of forage in *many* paddocks, you are probably overstocked.
- If you have misjudged quality, move immediately and note that on the chart.
- If you suspect you are over stocked, do a *field check.*
- To figure the size of sample areas, multiply the minimum grazing period (Col. 4) by the figure in row 34 and then divide by the size of the paddock (Col 3) just as we did for paddock A in the example above.
- If the sampled areas will not feed one animal for one day, you are overstocked and should destock immediately.

Step 17. Record Results.

The purpose for recording is to fine tune paddock assessments and management decisions in the future. Your plan becomes your permanent record. This applies to both open and closed plans.

- Record actual events *in ink* with a line in every paddock that covers the number of days actually spent there.
- Behind (or to the right) of that, record the volume of forage actually taken in ADA and follow it with an assessment of the grazing—"L" for light, "H" for heavy and "M" for moderate—i.e. if they took 34 ADA in a moderate grazing, record 34/M.
- Flag or mark any paddocks where serious errors were made—grazing period too long or recovery periods too short, etc.—that might affect the next plan.
- Record precipitation in rows 21 and 22. The spaces between the heavier lines are five-day periods. Note the exact day of major storms with a dot and a written comment in "remarks." Note average annual precipitation in row 37. Record the total for the season in row 38.
- Add any comments that will help in future planning on the back of the chart.
- Summarize livestock and land performance over the year in the lower right corner. The blank lines are for recording any other significant factors. The most important is total yield per acre of product sold.

Actual Grazing Record

YEAR 2005   PLAN Open-Ended Plan

ADA/H Actual/Estimate	Estimated Relative Paddock Quality	Paddocks Size	Number/Name	March	April	May	Paddock Number/Name
/		2000	A	8.7―2.3/L			A
/		850	B1		―2.4/L	7.3―	B1
/		870	B2			7.1―	B2
/		2200	C	―8.8―――――8.2/M			C
/		560	D1			5.5―	D1
/		250	D2				D2
/		370	D3			8.4―	D3
/		320	D4			9.6―	D4
/		2500	E1		―2.5/L 7.4―		E1
/		400	E2			5.1―	E2
/		750	F1		CONTINUOUS GRAZED		F1
/		1000	F2	3.0― .75/L			F2
/		200	H1				H1
/		450	H2	CROPS			H2
/		400	H3				H3
/							
/							
/							
/							
/							
		21. Rainfall		1.5	.75	2.0	
		22. Snow					

23. Growth Rate (F/S/0)	S S S S S F	F F		
24. Supplement or Feed—Type and Amount				
25. Number/Size of Herds	750	1030	1030	
26. Paddocks Available	12	12	12	
27. Recovery Period(s) or Number Selections	30-90	30-90	30-90	
28. Avg GP or AMGP/AMxGP	2.7-8.1	2.7-8.1	2.7-8.1	

Open-Ended Plan Only also record:

- Opinion of daily growth rates in the same 5 day periods on row 23 using "S" for slow and "F" for fast and "0" for no growth.
- When the growing season ends, draw a brown line down through all paddocks and label it "Growth Ended."

THE CLOSED PLAN

As with the Open-Ended Plan, follow along with the Sample Grazing Chart (Closed Plan) and Chart with Grazings Plotted (Closed Plan). Each consists of two pages.

Sample Grazing Chart (Closed Plan)

Holistic Management International

YEAR: 2005/06 PLAN: OPEN-ENDED / **CLOSED**

Grazing Plan
(Livestock/Wildlife)

Hunting Season

Columns 1 ADA/H Actual/Estimate	Columns 2 Estimated Relative Paddock Quality	Paddocks Size	Paddocks Number/Name	OCTOBER	NOVEMBER	DECEMBER	JANUARY
/		2000	A				
/		850	B1				
/		870	B2				
/		2200	C				
/		560	D1				
/		250	D2				
/		370	D3				
/		320	D4				
/		2500	E1				
/		400	E2				
/		750	F1				
/		1000	F2				
/		200	H1				
/		450	H2				
/		400	H3				
/							
/							
/							
/							

		OCTOBER	NOVEMBER	DECEMBER	JANUARY
21. Rainfall					
22. Snow					
23. Growth Rate (F/S/O)					
24. Supplement or Feed—Type and Amount					
25. Number/Size of Herds		1/1030	1/750	1/750	1/750
26. Paddocks Available		14	14	14	14
27. Recovery Period(s) or Number Selections		3	3	3	3
28. Avg GP or AMGP/AMxGP					

Type of Animals	No.	Average Weight	% Unit	Total Units	No.	Average Weight	% Unit	Total Units	No.	Average Weight	% Unit	Total Units	No.	Average Weight	% Unit
29.															
30.															
31.															
32.															
33.															
34. Total SAUs															

35. Cell Size: 13,120
36. Stocking Rate: 1:13
37. Avg Annual Precipitation:
38. Season Total Precipitation:

Remarks: Crops | Weaning | Continu...

an & Control Chart
dlife/Crops/Other Uses)

Grazing Cell: River Bend Ranch

FEBRUARY	★	MARCH	APRIL

Calving

COLUMNS							
3	4		5	6	7	8	
Paddock Number/Name	Non Growth	Drought Reserve	Non-Growing Season				
			Estimated Available ADA/H	ADs	Planned Demand ADA/H	Total Yield ADA/H	
A							
B1							
B2							
C							
D1							
D2							
D3							
D4							
E1							
E2							
F1							
F2							
H1							
H2							
H3							

1/750	1/750	1/825
14	12	12
3	3	3

Total Units	No.	Average Weight	% Unit	Total Units	No.	Average Weight	% Unit	Total Units	No.	Average Weight	% Unit	Total Units

tinuous Grazing   ★ Expected start of growth and start of drought reserve should growth not start

A. Estimated Total ADs (Livestock/Wildlife) _____
B. Estimated Days of Non Growth _____
C. Days of Bulk Feeding _____
D. Days of Drought Reserve Required _____
E. Total Days Grazing Required _____
F. Estimated Carrying Capacity _____

SUMMARY LIVESTOCK AND LAND PERFORMANCE
Calving/Lambing/Kidding _____ %
Avg Weaning Weight _____ Age _____ (Mths)
Daily Weight Gains _____ _____ Lbs
          (Growing)   (Non-Growing)

SDA(H)/inch (cc) rainfall _____
Total Yield (sold) per acre/hectare _____ Lbs

Chart With Grazings Plotted (Closed Plan)

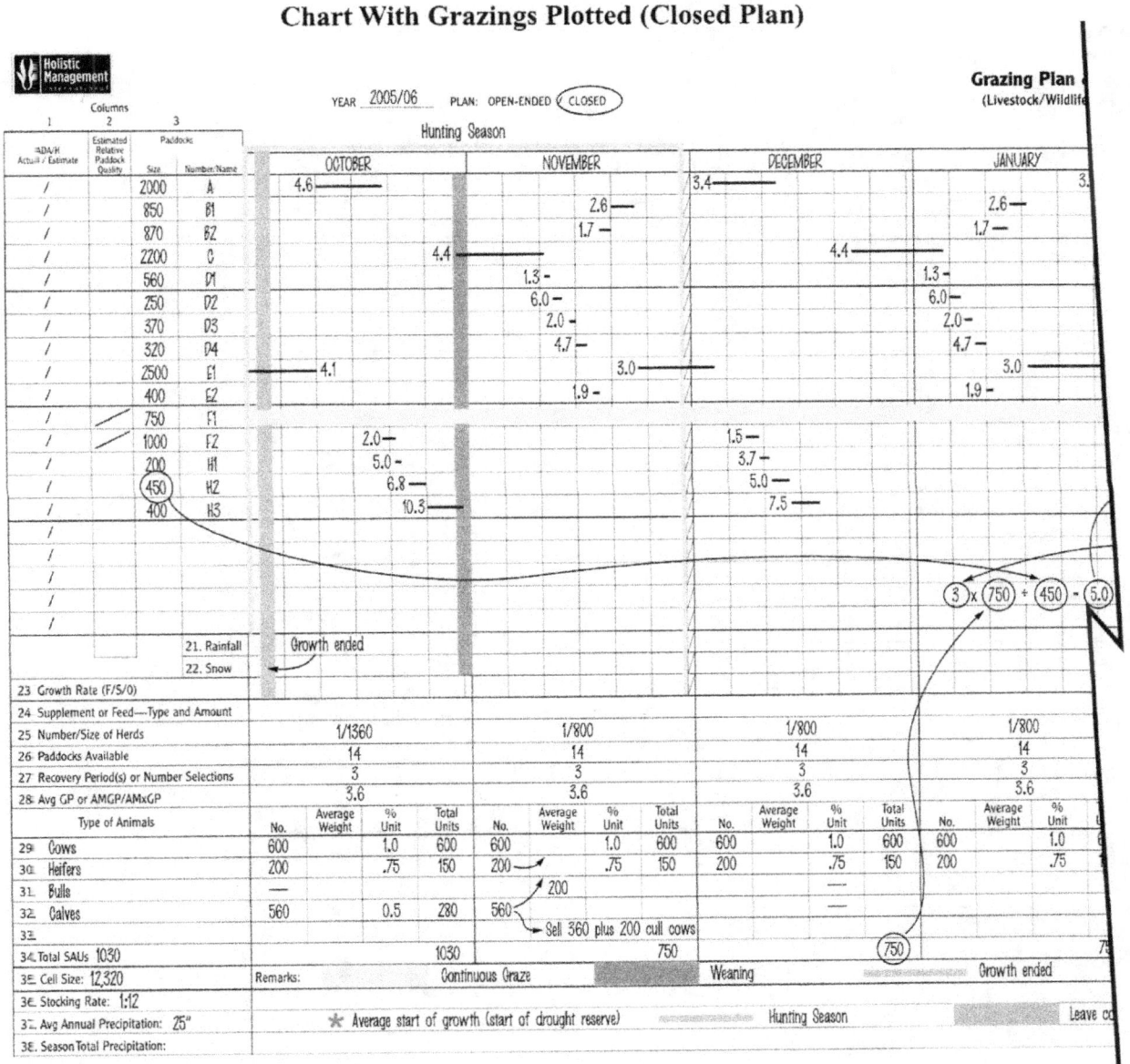

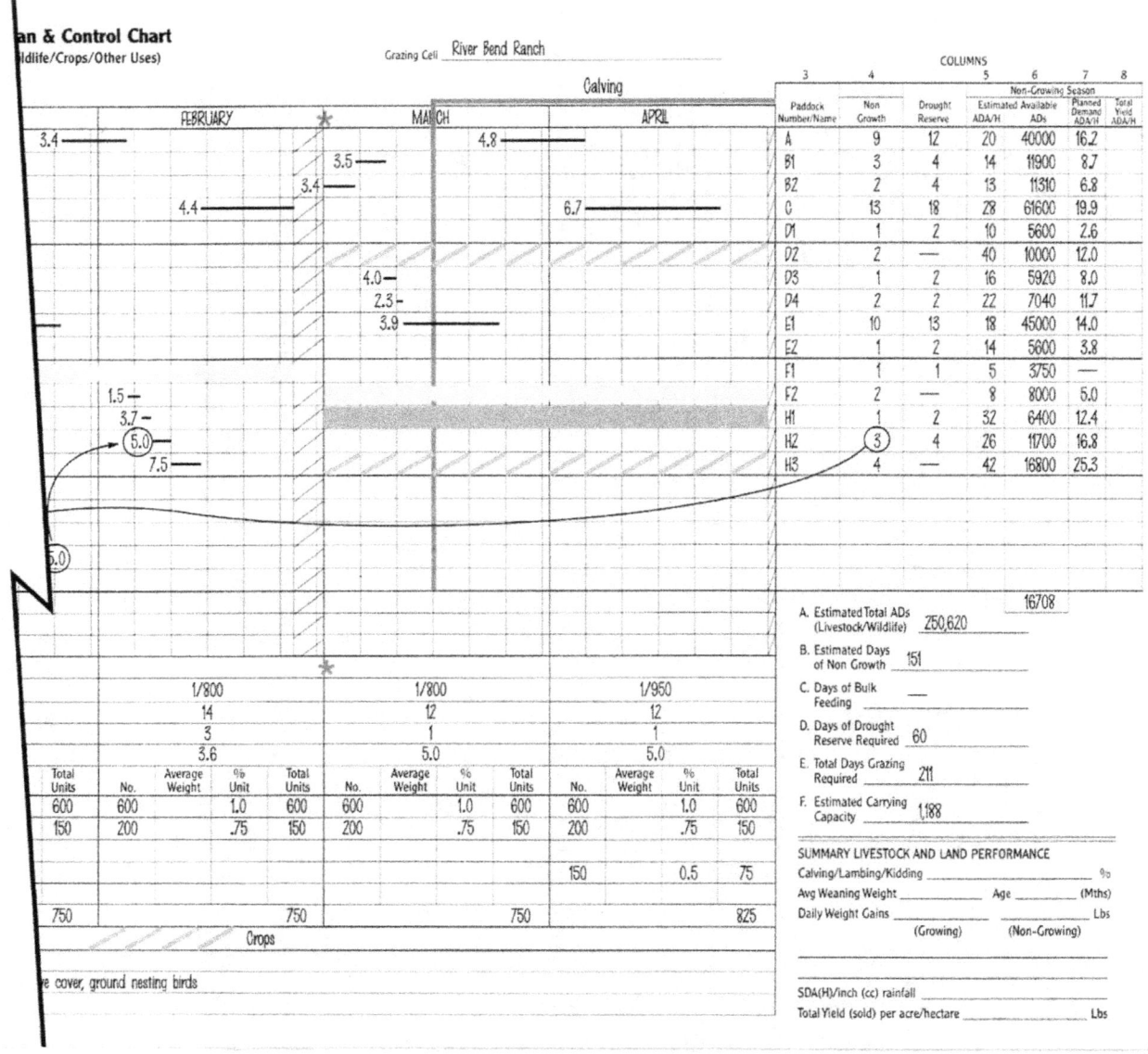

Step 1. Make Initial Decisions: Look at the Big Picture: Same as Open Ended Plan.

Step 2. Set Up the Grazing Chart: Same as Open Ended Plan with one addition: Closed plan only: Change the heading of column 4 to "Non-growth" and write "Drought reserve" in the blank column to the right.

Step 3. Record Management Concerns Affecting the Whole Cell: Same as Open Ended Plan with one addition: For a closed plan, mark the expected start of next season's growth which would also be the date you would have to begin using the drought reserve.

Step 4. Record Herd Information: Same as Open Ended Plan.

Step 5. Record Livestock Exclusion Periods: Same as Open Ended Plan.

Step 6. Not Applicable: Step 6 applies only to the Open-Ended Plan

Step 7. Record Paddocks Still Available: Same as Open Ended Plan

Step 8. Note Paddocks Requiring Special Attention: Same as Open Ended Plan.

Step 9. Rate Paddock Productivity: If this is your first *closed* plan you have no option but to start by estimating ADAs.

- *In each paddock*, pace off *several* random squares that could feed one animal for a day and compute the area in square yards.
- The problem essentially comes down to working the *field test* problem we did above backwards. Say for example: You find that the average square area of Paddock A that it will take to feed one of your animals for one day is a square of 24 yards on each side. Square that and you get 576 yds^2 for one of your animals for one day. Divide that into 4,840 yds^2/acre and you get 8.4 Animal Days per Acre (ADA) (as opposed to 8.7 before due to rounding). Repeat this for each paddock and record your results in column 5.
- Be pessimistic. Average the answers to get an estimate of ADAs. This is difficult if the terrain is varied and you are new. One way to allow for this is to remove all the acreage covered by the inaccessible areas. Also, you can sample within the poorer grassland areas more than the good ones. Record the figure to the right of the slash in column 5 (closed plan).

If this is *not* your first open plan, add up the ADA figures from column 8 in your last *closed plan*. They should appear in ink next to each grazing line. *Closed plan*: transfer them to column 5. Adjust them up or down based on how much better or worse production was in the just past growing season. See the Chart with Grazings Plotted (Closed Plan) column 5.

Factor Forage Quality and Area into Paddock Productivity: If any paddock is unavailable for the entire period (sacrificial paddock) do not include it in the ratings calculations

For Your Closed plan

1. Multiply the ADA in column 5 by the number of acres (col 3) in each paddock and record the result in column 6
2. Add column 6 and enter the total in row A. Divide row A by the total number of paddocks. Enter the result in the box at the bottom of column 6.

Here is how that segment of your grazing chart should now look: Also see the completed chart.

Rating Relative Paddock Quality (Closed Plan)

1	2	3		5	6	7	8
ADA/H Actual / Estimate	Estimated Relative Paddock Quality	Paddocks		Non-Growing Season			
		Size	Number/Name	Estimated Available ADA/H	ADs	Planned Demand ADA/H	Total Yield ADA/H
/		2000	A	20	40,000		
/		850	B1	14	11,900		
/		870	B2	13	11,310		
/		2200	C	28	61,600		
/		560	D1	10	5,600		
/		250	D2	40	10,000		
/		370	D3	16	5,920		
/		320	D4	22	7,040		
/		2500	E1	18	45,000		
/		400	E2	14	5,600		
/		750	F1	5	3,750		
/		1000	F2	8	8,000		
/		200	H1	32	6,400		
/		450	H2	26	11,700		
/		400	H3	42	16,800		
/							
/							
/							
/							
/							
		21. Rainfall					
		22. Snow					

250 x 40 = 10,000

Average ADs → 16,708

Step 10. Not applicable- Open-Ended Plan Only

Step 11. Not Applicable - Open-Ended Plan Only

Step 12. Closed Plan Only: Assess Forage Volume, Carrying Capacity and Drought Reserve

Will you have enough forage to carry you through the non-growing season and provide a drought reserve? Use the rows on the lower right of the chart to record:

- *Expected days of non-growth.* Enter in row B
- *Days of bulk feeding.* (Due to heavy snow or something similar) in row C. Determine the type and amount of feed required and enter under the appropriate month in row 24.

- *Total days grazing required.* Add rows B and D and subtract row C. The result is the total days of grazing required from the land. Record in row E.
- *Estimated carrying capacity.* Divide row A by row E. This is the number of animals the land can support without additional bulk feeding. Record in row F.

If row F is much lower than the number you have now, reduce animal numbers now. The earlier you do it the less you will have to do it.

Step 13: Closed Plan Only: Plan the Number of Selections and the Grazing Periods.

It is difficult to overgraze plants that are not growing so recovery periods are less critical. You just want to keep stock concentrated enough for impact and give paddocks enough time for fouling to wear off. The focus is on rationing the forage through the non-growing season and drought reserve while trying to keep the plane of nutrition as even as possible.

Every time animals return to a paddock, forage has decreased in volume and quality. So, the fewer times a paddock is grazed, the better. But, as the grazing period is lengthened, animals become nutritionally stressed the longer they stay in a paddock. Also, it can be detrimental to shorten the grazing periods to the extent that recovery is insufficient for fouling to wear off. In other words, you have some thinking to do.

Number of Selections and Average Grazing Period: The ability to minimize the number of times animals graze a paddock depends on the number of paddocks which, in turn, influences the recovery time and the average grazing periods. Here is how to figure it out:

- Divide total days of grazing required in the non-growth period (row B) by the number of selections you want to use (try for one or two) to see what the recovery period would be. For example, if total days of grazing = 211 and you want 2 selections, 211/2 = 106 days of recovery from fouling between grazings. Then…
- Divide the recovery period by the number of paddocks to get the average grazing period —e.g. if you have 14 paddocks, then 106 days/14 paddocks = 7.57 days/paddock.

So, with 14 paddocks, 2 selections of 8 days each will cause forage quality to decline to the extent that you may need to supplement from the very start of the first selection.

Refer to the Plotted Chart: But, with 3 selections, you would have 70 days of recovery (row E divided by 3) (211/3) which is adequate for fouling to wear off. Paddocks grazed three times for five days will reduce the need for supplement in the first selection. The grazing being taken more evenly will delay nutritional stress on the animals. You might have to supplement during the second selection and almost certainly during the third, but overall, you will need less supplement.

On the other hand (and especially in the early years when you have few paddocks): One selection might be best. Stock density will automatically be low and thus fouling is not the big factor that it will be as paddock numbers increase.

Record the number of selections in row 27 and the average grazing period on row 28.

Use a single selection if you have a great many paddocks because grazing periods are so short (one or two days) that animals do not remain long enough to be nutritionally stressed. For example: With 180 days of non-growth, recovery period = 180 days/1 selection = 180 days recovery. 180 days recovery/100 paddocks = 1.8 days (180 days/99) average grazing period. Keep at least one decimal place in row 28.

Drought Reserve: Use the same procedure to calculate the average grazing period for the drought reserve—e.g. if row D = 60 and you had 14 paddocks, the average grazing period would be 60/14 = 4.3 days with one selection. Enter in row 28 under the appropriate weeks or months. For example: The Plotted Chart indicates that, on the average, the growing season begins about the first of March. After that, one 5 day selection in each of the 12 available paddocks would take care of the 60 day drought reserve.

Actual Grazing Periods: Convert average grazing periods (row 28) into actual using the paddock ratings in column 6 and using the figure at the bottom as your average. Record in column 4.

$$\text{Grazing period} = \frac{\text{Paddock rating} \times \text{Average grazing period}}{\text{Average paddock rating}}$$

Example: Paddock A: Paddock rating = 40,000 (column 6); Average grazing period = 3.6 (row 28); average paddock rating = 16,708 (bottom of column 6): So, 40,000 X 3.6 = 144,000 ÷ 16,708 = 8.62 rounded to 9 and entered in column 4.

Use the same procedure for the drought reserve (40,000 X 5 day grazed period)/16,708 = 11.97 rounded to 12 and entered in column 4 under drought reserve.

If paddock numbers change (check row 26), do a second calculation being sure to adjust average paddock rating to reflect only the paddocks to be used. Record this beside the first figure in column 4 separating the two with a slash.

Step 14. *Plot the Grazings*: Same as Open-Ended Plan with one addition: *Closed Plan*: Plot the moves from where you are currently to when you anticipate the next growing season to begin plus an additional drought reserve. You have a fixed amount of forage to ration. If growth starts when you expect, all is well. But, you will have some old drought reserve forage to deal with.

Step 15. *Make a Final Check of Your Plan*: Same as Open-Ended Plan with one addition: *Closed Plan*: Add all the ADAs to the left of the grazing line and record what you get in column 7. Then…

Make a field check. Again, divide 4,840 by the ADA figure in column 7 for all paddocks and take the square root of the result. Pace out the square and ask, "Will this area feed one animal adequately and *still leave enough litter and feed for wildlife?*"

Refer to column 7 of the Closed Plan with Grazings Plotted. The total ADA planned for paddock A is 16.2, for B1 it is 8.7 and so forth. The total of column 7 is 163.2 ADA for the whole cell. Divide that into 4,840 and you get 29.7 the square root of which is 5.4 so to be conservative, use 6 yards. Pace off representative sample square areas of six yards on each side and test with the questions.

Step 16. *Implement (and Monitor) the Plan*: Same as Open-Ended Plan with additions:

Closed Plan Guidelines:

- *Monitor forage consumption.* If you run out of forage in any paddock, you probably misjudged volume or quality. Note that on the chart.
- If you run out in many paddocks, you are probably overstocked
- If you are concerned for the health and stability of wildlife, monitor to ensure cover, feed, lack of disturbance, etc. are being allowed for.

If you Experience a Drought: If the growing season does not start when expected, begin using your drought reserve. Although you estimated this conservatively, reassess your plan to take into account actual animal days used during non-growth and adjust the remaining available ADs. Where necessary, recalculate the grazing period for each paddock. You may get some greening up even without rain. Although this improves nutrition, these plants are very susceptible to overgrazing.

Drought Guidelines:

- Do *not* speed up moves to chase the green. Growing plants require adequate recovery time. Simple rule:
    - For every day animals stay in a paddock longer than they need to, their performance will decline.
    - But, for every day they move out sooner than they need to, you lose a day of recovery time in every paddock.
- Failure to understand this cumulative effect on recovery leads to managers favoring animals to the point that they run out of forage and then they blame it on the "drought."
- *If the rains never materialize and you run through your drought reserve time, create a new closed plan ASAP.* (The same holds if you get early rains but nothing more.) *Assume you will not get any rain until the next growing season and plan right on through until then. Now run field checks for the new plan which are likely to show that you have too many animals. Destock now*—the earlier you destock the less you have to destock. If you are not already running one herd, seriously consider doing it. Keep animals concentrated and moving constantly.
- *When the rains do start* create an open ended plan. If this happens before you are through your drought reserve, you will have some old forage to remove as soon as you can.
- *In low, erratic rainfall areas, plan drought reserves of 6 to 12 months.* Start a new grazing chart (closed plan) once you run through the start of anticipated growth. Since you were planning a prolonged drought reserve, your stocking rate may be relatively low. However, field checks of the new closed plan may show that you have too many animals. If so, destock now—the sooner you destock the less you will have to destock.

Thoughtless rotation is a major cause of stock stress in the non-growing season

Step 17. Record Results. Same as Open-Ended Plan with additions:

Closed Plan Only also record:

- If there is any growth over the planned period, put the letter "G" in row 23. Some greening up occurs in many areas in the non-growing season. This may spread over more time each year with an increasingly effective water cycle.
- When the next growing season begins draw a green line down through all paddocks for that day and label it "Growth Started." At this point, create you next open-ended plan.
- Record the forage each paddock yielded in the growing season. Add all the grazings in ADA from the green line to the brown line. Record this in column 8 with a light pencil.
- You now need to make some adjustments to these figures:
    - Paddocks that were grazed late last growing season (and thus did not fully re-grow) actually yielded more than the penciled in figure shows.
    - Also, any paddocks that have an "M" or "L" behind the last grazing taken had more forage than the ADA taken figure shows.

- o Estimate the additional forage and add that to the yields in column 8. Record the final figures in ink. These represent the best possible comparison of paddock quality and "performance."

A special case: Grazing two or more herds in the same cell

There are 3 ways you can run two or more herds in the same cell:

- Move separate herds among all paddocks while keeping recovery times adequate.
- Allocate certain paddocks to each herd and plan each division as a sub-cell.
- Have one herd enter a paddock as another leaves (follow-through grazing).

None of these should be attempted if you have only a few or very uneven paddocks because grazing periods will be unacceptably long.

You might choose the first option if you want to give a smaller herd (e.g. steers you plan to sell) a larger area to select from. The larger herd will go through those same paddocks at higher density.

You might choose the second option if you have some animals that you want to run in a few paddocks close to the house or barn. The main herd would move through the rest of the paddocks.

Follow-through grazing is tricky to plan. It might be the choice if you have herds that have different nutrition requirements—i.e. first calf heifers vs. mature cows or high-performing dairy milkers vs. non-milkers.

The Basics: The figure below shows two follow-through cases:

Follow-Through Grazing Planning

1	2	3	4	5	6	7
180 acres	180 acres	180 acres	180 acres	180 acres	180 acres	180 acres
12 Days	12 Days	12 Days	12 Days	12 Days	12 Days	12 Days

1	2	3	4
140 acres	140 acres	140 acres	280 acres
	15 Days	15 Days	30 Days
7	**6**	**5**	
15 Days	140 acres	140 acres	280 acres
	15 Days	15 Days	30 Days

Cell I: Follow-through grazing with 12 days in equal paddocks gives 60 days recovery.

Cell II: The follow-through plan below gives 60 - 105 days recovery in unequal paddocks.

A grazing plan for 60-day recovery times in Cell II might look like this.
Note paddocks 3 and 6 are only grazed once, and 4 and 5 are grazed too long.

Herd 1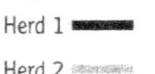
Herd 2

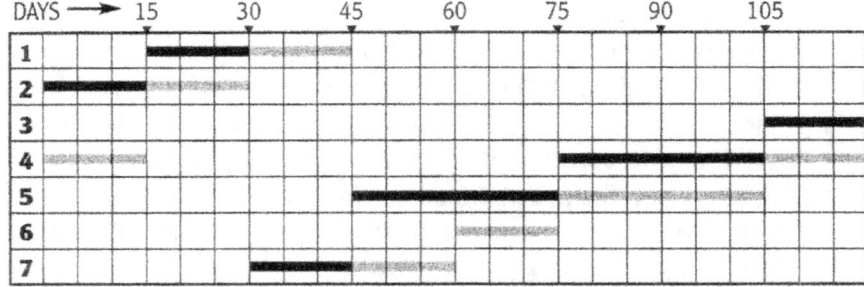

In Cell I, although plants are exposed to animals for twice as long as they would be with one herd, equal paddocks are no problem mathematically. But, two herds in Cell II with unequal paddocks complicate the math. Use the following guidelines while carefully and continually monitoring.

- If one herd is small, key all moves to the larger herd.
- If the moves progress from larger to smaller paddocks, the following herd can skip a paddock occasionally in order to catch up.
- If moves progress from smaller to larger paddocks, the lead herd can skip ahead while the following herd actually makes the first grazing of the intervening smaller paddocks.

Note: when one herd skips a paddock, you shorten recovery times. So try to plan similar paddocks in sequence.

Creating Your Plan

Before you decide to run more than one herd, think through the consequences. Record your decision on the Grazing Chart in row 25. For two herds of 100 and 500 head, enter 2/100-500 and note which herd goes first. For example, record 50 replacement heifers followed by 250 cows as 1/50h-250c. Use whatever abbreviations are clearest to you.

Calculating Average Grazing Periods

In Step 11 (open plan) and Step 13 (closed plan) use the following formulas:

Two or More Herds Using Any Paddock in the Cell

Open-ended Plan:

$$\frac{\text{Minimum recovery period}}{(\text{\# paddocks} \div \text{number of herds}) - 1} = \text{Average Minimum Grazing Period}$$

$$\frac{\text{Maximum recovery period}}{(\text{\# paddocks} \div \text{number of herds}) - 1} = \text{Average Maximum Grazing Period}$$

Closed Plan:

$$\text{Average Graze Period} = \frac{\text{\# drought reserve days, (or \# non-growing days)} \div (\text{\# paddocks} \div \text{\# herds})}{\text{\# selections}}$$

Example: If you have a 180 day non growing season, 50 paddocks, 2 herds and two selections, your average graze period would be 3.6 days. Record this in row 28 under each month (e.g. 2.9 – 11.7 open; or 3.6 closed).

Two or More Herds with Certain Paddocks Allocated to Each Herd

Open ended Plan: Calculate four average grazing periods—two per herd.

Herd one:

Minimum recovery period
$$\frac{\text{Minimum recovery period}}{(\text{\# paddocks allocated } -1)} = \text{Average Minimum Grazing Period}$$

Maximum recovery period
$$\frac{\text{Maximum recovery period}}{(\text{\# paddocks allocated } -1)} = \text{Average Maximum Grazing Period}$$

Herd two: Repeat the above using the paddocks allocated to the second herd

Record all four in row 28 (e.g., 3.6-11.7 / 2.5-6.6). Go to column 3 and color-code the paddocks to be used by each herd. This will be needed in calculating actual grazing periods.

Closed Plan: You must calculate two average grazing periods—one per herd for both the non-growing months and the drought reserve period.

Herd one:

$$\text{Average Grazing Period} = \frac{\text{\# non-growing days, or \# drought reserve days} \div \text{\# paddocks allocated}}{\text{\# selections}}$$

Herd two: Repeat the above using the paddocks allocated to the second herd. *Example*: Given a non-growing season of 180 days, 100 paddocks and two selections with 25 paddocks to heifers and 75 paddocks to cows: The average grazing period = 3.6 days for heifers and 1.2 days for cows.

Record both figures under each month in row 28 in both the non-growing period and drought reserve (e.g. h/3.6—c/1.2). Color-code which paddocks are to be used by each herd in Column 3. This will be needed in Steps 12 and 13.

Two or More Herds on Follow-through Grazing

Open Ended Plan: If you are using one recovery period, calculate one grazing period which will be used by each herd. For example, say you have 2 herds and a 2 day grazing period. The first herd will be in the paddock for 2 days and the second for the following 2 days. So, the paddock would have animals in it for 4 days (2 herds X 2 days).

$$\frac{\text{Recovery period}}{(\text{\# paddocks - \# herds})} = \text{Average Grazing Period}$$

If you are using a range of recovery periods, calculate two average grazing periods—the shorter for fast growth and the longer for poor growing conditions. This is very difficult to manage if paddocks are uneven in size or quality.

Minimum recovery period
$$\frac{\text{Minimum recovery period}}{(\text{\# paddocks - \# herds})} = \text{Average Minimum Grazing Period (for each herd)}$$

Maximum recovery period
$$\frac{\text{Maximum recovery period}}{(\text{\# paddocks - \# herds})} = \text{Average Maximum Grazing Period (for each herd)}$$

Record these on row 28 (e.g. 3.5 or 1.2-4.3)

Closed Plan: Calculate only one grazing period but it will be used by each herd. For example, with 2 herds and a 2 day grazed period, the first herd would be in each paddock two days (average) and then the second herd would be in it for the next 2 days. So, the paddock would have animals in it for four days. One selection is the norm with follow-through grazing.

$$\frac{\text{\# non-growing days or \# drought reserve days}}{\text{\# paddocks}} = \text{Average Grazing Period (for each herd)}$$

Example: Given: a 180 day non-growing season, 100 paddocks, 2 herds on follow-through. Each herd would spend an average of 1.8 days in each paddock. Record both figures under each month in row 28 (e.g. 1.8—1.8)

Calculating Actual Grazing Periods

Factor in paddock productivity as in Step 11 (open plan) and Step 13 (closed plan).

$$\frac{\text{Paddock rating X Average Grazing period}}{\text{Average paddock rating}} = \text{Actual Grazing Period}$$

Distinguishing One Herd from Another

In step 14, use symbols that will allow you to distinguish one herd from another and note which symbol represents which herd in the "remarks" section.

A More Refined Method for Calculating Grazing Period Adjustments Based on Herd Size Changes

When creating a closed plan and herd size varies dramatically, consider a more refined method for adjusting the grazing periods than that suggested in Step 15 above.

Start by determining a herd size adjustment factor based on the difference between average and actual. Then adjust the grazing periods in Step 14 based on the adjustment factor. Follow these steps:

Step 1: Calculate the average herd size for the non-growing months and then for the drought reserve.

- Multiply SAUs (row 34) by the number of days the herd stays the same size. Record the AD in brackets on row 34. When herd size changes, do the same again and so forth until the end of the planned period (non-growth or drought reserve).
- Add up all the AD figures and divide by the total days in the planned period. Record this "average herd size" in the "remarks" section.

Example: Non-growing Months: Assume the planned period is 200 days. Herd size during the first 60 is 450 SAU (450 X 60 = 27,000 ADs); then 300 SAU for 45 days (300 X 45 = 13,500 AD); then 375 SAU for 95 days (35,625 ADs). Then calculate the average:

$$\frac{27{,}000 + 13{,}500 + 35{,}625}{60 + 45 + 95} = \frac{76{,}125}{200} = 381 \text{ SAU Average Herd Size}$$

Example Drought Reserve Period: Assume 60 days. Herd sizes: 375 SAU for 15 days; 450 SAU for 30 days; then 500 SAU for 15 days. Again, multiply SAUs by days to get ADs, then add the ADs and divide the result by total days. You should get an average herd size of 444 SAU.

Step 2: Calculate an adjustment factor for each herd size by dividing the average by the actual size.

$$\frac{\text{Average herd size}}{\text{Actual herd size}} = \text{Adjustment Factor}$$

Example: Non-growing Months. Continuing the above example: Average herd size = 381. First 60 days, actual = 450, so adjustment factor = 381 ÷ 450 = 0.9; next 45 days, actual = 300, factor = 381 ÷ 300 = 1.3; last 95 days, actual = 375, factor = 381 ÷ 375 = 1 (no adjustment needed).

Example: Drought Reserve Period: Average herd size = 444 SAU. First 15 days, actual = 375. So, 444 ÷ 375 = 1.2 adjustment factor; next 30 days, actual = 450 SAU, 444 ÷ 450 = 1 (no adjustment needed); last 15 days, actual = 500, so 500 ÷ 444 = 0.9.

Step 3: Record adjustment factors in brackets to the right of herd size in row 25. Then multiply each of those figures by the actual grazing period in column 4 (or the drought reserve to its right). Round off to the nearest whole day (or half day) and pencil in the adjusted grazing period line.

Occasionally, herd size will change during a planned grazing period in an individual paddock. The following procedure should be used (especially if herd size changes dramatically and the calculated actual grazing period is very long).

Step 4: Calculate how many ADs would be harvested based on the average herd size.

For example: assume the actual grazing period in column 4 is 25 days with an average herd size of 300. Thus you would have 7,500 ADs.

Step 5: Now calculate how many ADs the actual herd will have accumulated up to the time the size changes.

For example, say herd size = 200 SAU during the first 10 days. Thus, ADs harvested would be 2,000. That leaves 5,500 ADs (7,500 – 2,000). So, if herd size after day ten increase to 280 SAU, then there are 20 days of grazing left in the paddock (5,500 ÷ 280 SAU). So, the grazing period can be extended from 25 to 30 days (10 + 20) which is penciled in on the chart.

Planned Grazing Summary

Each step builds on the previous one so the order is important. Don't skip any steps.

Of all the procedures, grazing planning is the most orderly even though its object (biology) is the most unpredictable, complex and baffling subject that human intelligence every tried to fathom.

Living organisms never stop changing. Thus, you can never stop planning. Doing so will eventually lead to ruin.

You plan for living things, not because you can ever hope to bind them to your pattern, but so you can fit into theirs.

Lesson 8: Financial Planning and Monitoring

Introduction and Overview of the Planning Process

If prosperity and financial security are included in your goal, financial planning is the single most important activity undertaken each year.

The following is a summary and synopsis of the financial planning chapters in *Holistic Management* and its companion *Handbook*.

Land & Livestock International, Inc. uses an approach to financial planning and control that is substantially different and a great deal simpler, more effective and responsive than that presented here.

The Land & Livestock model takes advantage of modern computer technology to handle the mundane calculations while still requiring the manager to think through (sometimes daily) every single event that will take place on the ranch for the next twelve months.

It includes a perpetual 12 month rolling cash flow plan with weekly ending and beginning balances that provides instant feedback of budget and actual differences and projected balances every time a revenue or expenditure is recorded (preferably daily).

This cash flow plan is then integrated into a computerized (by enterprise) direct cost accounting system that has a budgeting feature. Financial statements are automatically updated after each transaction is posted and presented showing actual vs. budgeted amounts and variances (differences).

We do, however, follow HRMs general philosophy as to planning profit first and prioritizing expenditures during the budgeting process. This is a key element in financial success.

According to the "holistic" approach, conventional financial planning usually has three stages:

1. An estimate of income from the enterprises determined to be the most profitable.

2. Expenses are then budged in columns for capital investment, overheads, variable costs, loan repayments, etc.

3. Finally you calculate the cost of borrowed money and the bottom line emerges as a profit or a loss.

So far, we are fine with that. But, they erroneously continue by claiming that, as long as the plan "cash flows" there should be no problem and all is considered to be well. That is not necessarily so.

They also claim that two things distinguish the kind of financial planning they do from the conventional:

1. It is guided by the "holistic" goal. That does not mean that "conventional" financial planning is not guided by the entity's "strategic" goal. And...

2. It accounts for human nature and requires a determined attitude toward success. Where is it written that the "conventional" approach does not do this as well? Nowhere.

General Planning Guidelines, Management Tips and Good Ideas

Psychology of Planning: People's attitudes cause them to plan unprofitability and then complain loudly when the plan succeeds. The underlying tendency for all of us is to allow expenses to rise to meet the income we anticipate receiving.

The Debt Trap: In agriculture, income comes only once or twice a year. Thus, the tendency to let expenses rise to anticipated income leads to trouble. For example, what if calf or lamb prices suddenly crash and you are operating on borrowed money? Servicing that debt becomes a major outlay which is an unproductive expenditure because it will not generate any additional income.

Plan Profit before you Plan Expenses: Planning profit *before* allocating money to expenses will counter the tendency to allow costs to rise to anticipated income. Once you have figures for total expected revenue, cut it by one half and set that much aside as profit. This needs to be a substantial amount in order to place severe restraints on the funds left for operating the business. If it is too low, there is no challenge to keep costs down. On the other hand…being too high is demoralizing. *Planning profit first is an essential exercise for prompting the collective creativity and concentrated effort necessary to produce amazing results.*

Early on in the process you should determine who should be involved and to what extent: Sometimes you may need the skills of an accountant but accountants should never be allowed to lead the planning process. The leader should have a good grasp of the business and the skills to facilitate and coordinate the process. In larger organizations, responsibilities should be divided among teams with each enterprise having a team. Expenses that are not linked directly to a particular enterprise (overheads) need to be assigned to one or more people or teams. The main planning team should include representatives from each management team. The people who are responsible are the ones that need to come up with the figures that are under their control. *Without ownership in the figures, there is no incentive to make sure they stay on track.* Since everyone is working toward the same strategic goal, they need to feel assured that the plan is taking them toward that goal. Finally, the main planning team needs to be limited in size or the process will bog down.

Rewarding your People's Managerial Effectiveness: The environment that motivates people is the one where ideas and efforts are recognized and where they are genuinely involved in decisions. The best financial reward program is one that is fair and does not destroy trust.

Conventional productivity based bonus programs usually fail on both counts for good reason. Bonuses based on production, income or profits have very little potential for inspiring people. Such things are usually so remote from daily decision making that workers never really know where they fit into the picture. If managers push up production or income, they usually also push up costs. If input costs increase along with output (which is highly likely) the business may be actually running at a loss. Regardless, you are still obligated to pay the bonuses.
Also, it can give rise to mistrust as to how the profit was determined.

It would be better to *base rewards on net marginal income derived by subtracting all the expenses under the control of those running the enterprise from all the income they produce.* This would pressure managers to both increase income as well as decrease costs. Only income and expenses under direct control of a management sector would be taken into account. True "open book" management would be required with all parties having access to the figures.

Also, it would essential to use some formula to account for "ownership" (fixed) costs off the top. Otherwise, it would break the owner. In other words, capital contributions of owners and bank loans would not be considered when calculating bonuses.

Simple Appreciation: How to Handle Emergencies: There are situations that cannot be accounted for during routine planning. Things will go wrong even with good planning. So, what we need is a way to handle such situations without clouded judgment due to stress. There is a technique that comes from the British military that can do exactly that. It is called "simple appreciation."

Take a deep breath to ease the panic. Then take out a note pad and start writing under four headings.

- **Aim.** Define your objective in one sentence. What result do you want?
- **Factors.** List everything, good or bad, that has a bearing on the situation. You will not look at what you write again so do it quickly and do not be judgmental. This is intended only to get what is in your head out into the open.
- **Courses.** Outline the options that would achieve your aim. (Rarely more than three.)
- **Plan.** Pick the course most likely to achieve the aim and plan whatever you need to do to take it.

Tracking Your Decisions: After planning for a year, the assumptions you used may not apply several months later when it is time to implement them. Recording these decisions to be made in chronological order can serve to warn you that a change may be needed before you implement the decision. Also, such a record helps minimize argument when team members have different recollections

(Editor's Note: Most of the forms referred to in this lesson are available at Holistic Management International's web site. Those that are not can easily be adapted from an old fashioned accountant's spreadsheet. Also, many of the modern accounting programs [Quick Books, Simply Accounting, etc] have features for creating budgets, calculating variances and presenting them as a part of the financial statements.)

Brainstorming: Generating New Ideas: The informal ritual of brainstorming is embedded in most planning procedures and is designed to open the minds of planners to new ideas. The most original and fruitful thinking occurs in an atmosphere of humor and playful competition.

Simple rules:

1. Get everyone who has an interest (and some who don't) together.
2. Divide them into groups and appoint a recorder for each group.
3. Conduct a short (about 3 minute) competition for the longest list of ways to solve some silly problem—What uses can you find for….? How would you deliver a proposal of marriage to…? No discussion allowed. Only numbers count.
4. At the end of the time, read the lists and award a humorous prize.
5. Now put the serious problem on the table. Allow ten minutes. Record any ideas that pop into their heads, no matter how ridiculous.
6. Appoint a facilitator to monitor: Do not allow judgments or conversations. Allow cheating—an idea picked up from another table can start a fresh run.
7. Pick out the ideas that have potential and develop them.

The Planning Process Itself

The process has two phases the first of which is *preliminary-information gathering* and decision testing done in several sessions held periodically over several months. It should be started well before the end of the fiscal year—3 to 6 months before the final planning session. Then the *second phase is to create the actual plan.* This is where you finalize revenue and expense figures and make any adjustments needed for a positive cash flow. This should only take a day or two.

Preliminary Planning Sessions: The preliminary planning sessions outlined below are usually necessary. But, some organizations may find that combining sessions or rearranging content is better. The following is only a guide:

First Session: Annual Review: This is the time where you need to consider where you have been and what might be preventing you from getting to where you want to be.

To address this, you'll need to consider three basic questions:

1. Is there a logjam? Mentally scan through the operation, find the blockage and determine the cause. The first place to look is within the organization—how committed are the people to the strategic goal? If commitment is lacking it may be that the goal is not clearly enough defined. Or, it could be that knowledge or skills are lacking or communication is poor. Market factors are seldom the cause but insufficient capital sometimes is. *If you find a logjam that takes money to remove, it will take priority when allocating money for expenses.*

2. Are there any other factors adversely affecting the business as a whole? Your accountant's workload may have increased to the point of needing an assistant. Maybe the old copy machine is constantly breaking down. Maybe you have outgrown your office and need to build or rent a bigger one. Such expenses would rank high on your priority list provided the need is genuine.

3. Are the gross profits on current enterprises as good as planned? Since you have been monitoring these monthly, you should be able to answer some simple questions. Are the planned amounts going to be achieved this year? Are they likely to increase, decrease or remain the same next year? If the plan is not likely to be achieved, why not? Unless it is a new enterprise where the learning curve has thrown your calculations off, either modify the product or drop the enterprise altogether.

Optional Session: Brainstorming New Sources of Income: Enterprises need to be challenged periodically but *don't overdo it. Few businesses survive by changing enterprises every year.* New enterprises involve a learning curve and may take a while to prove themselves. Brainstorming every 3 to 5 years is usually often enough.

Before adopting a new enterprise, consider:

1. They take time to develop. There is always a learning curve and it can be costly. They should solidly overlap old enterprises when possible.

2. A person's managerial effectiveness is diluted by the number of enterprises he is responsible for. Being stretched too thin may destabilize all enterprises.

3. *There is a direct relationship between management effectiveness and the proximity to what is being managed.* The more frequent your contact with an enterprise, the better you are able to spot problems early.

4. Often it is easier to modify or find new uses for an existing product than it is to create a totally new one. Most inventions occur by finding new uses for, or new configurations of, old products.

Selecting the Appropriate Enterprise: You must narrow down your list to the ones that suit you best.

First: *Drop any ideas that conflict with the society and culture test.* (The society and culture tests are hardly necessary. If you understand economics, you realize that profitability is the most efficient measure of success while simultaneously taking into account the social and cultural aspects.)

Second: *Determine the Enterprises you will engage in this year:* Drop those that are patently ridiculous but be careful. The best ideas are often ridiculed by the most knowledgeable people. Look twice. Before you reject them outright consider that they could spark ideas that are not so ridiculous.

Perform a detailed gross profit analysis of all the enterprises you might engage in. Include your current as well as any new ones. Gross Profit (revenue minus direct costs) is what each enterprise will contribute to the covering of overhead (fixed) costs.

If a new enterprise requires substantial capital investment, determine how that capital will be raised and repaid which will be needed in planning cash flow.

Even if you are not considering any new enterprises, this session is still essential because the gross profit on current enterprises will likely change with time. You will likely want to continue old enterprises that are sound.

But, if one of the new ones promises far greater returns, consider how to accomplish the changeover without loss or disruption.

Third: *Determine the Weak Link in Each Enterprise and How to Address It:* Conduct an informal gross profit analysis (you won't have actual figures to use at this point in the process). Keep the enterprises that bring in the most revenue for the least cost. Look for major differences, not minor ones. Single out those that clearly offer far more return than others for more detailed scrutiny.

The chain of production has three links (resource conversion [the land], product conversion [the livestock] and marketing). One of these will always be the weakest. When you allocate money, the weak link will have priority.

Once each team determines the weak link in its enterprise, consider what you need to do to address it. Each team reports its results to the main planning team. This is the reason for the 3rd session which is mostly for information sharing. Insights as outsiders may be obtained from people not directly engaged in a specific enterprise. Only rarely would the main planning team determine that another link was weakest because they seldom know the enterprise as well as those directly responsible for it.

Fourth: *Make a Preliminary Allocation of Expenses and apply the rest of the testing guidelines.* If the enterprise will lead you away from your strategic goal, eliminate it for now.

Considering the many expenses involved, it is easier if they are lumped into categories:

Wealth-generating expenses are those that increase your income over what you are currently earning. They *include the expenditures that address the weak link in each enterprise, clear logjams or rectify administrative shortcomings.* They include the expenses that address the upcoming weak link in the wealth-generating category. You will later test these against each other and may drop some of them in favor of those that produce the most new income.

Inescapable expenses are those that absolutely cannot be adjusted, delayed or changed in any way. Examples are payments on the contract to buy the business and land taxes.

Maintenance expenses are essential to running the business and maintaining present income levels but will not, themselves, generate any additional income. Most expenses fall into this category—supplies, utilities, telephone, etc.

You need fairly accurate figures for these expenses. This is data that the teams responsible for each enterprise should gather. The team responsible for overheads should gather the rest.

Wealth generating expenses should have money allocated to them first. First divide them into two groups:

1. Those that must have a specific amount of money or nothing will happen (for example, if it takes $3,000 for a new computer it is pointless to allocate $1,500). Relatively few expenses fall into this group and they are usually small. Sometimes, wealth generating expenses in this group might include the first payment on an item that may continue for several years. In subsequent years, these would be maintenance expenses.

2. Those that need every dollar they can get, but can still generate additional income with whatever you allocate to them (for example, advertising expenditures for a new product could be almost any amount).

Then second, apply the marginal reaction test to compare the wealth-generating expenses within each enterprise with an eye for eliminating expenses that provide small marginal returns relative to other enterprises.

If the first all or nothing group looks good, allocate money to them right away and then turn your attention to the other wealth-generating expenses. Particularly in livestock enterprises, one of these income generating expenses may produce such a high return that it would be senseless to invest in any of the others—e.g. compare fence building to brush clearing.

In other cases, you commonly need to allocate a minimal amount to several wealth-generating expenses in the second category because their return is spread fairly evenly between them. If the problem is serious, allocate money to it and then turn to the proposed actions that require a minimum investment but need all the money they can get—for example, improving internal roads.

Wealth-generating expenses are important. But, you still have to have money for maintenance expenses to maintain present income levels and cover any inescapable expenses.

Each of the responsible teams summarizes their results for the main planning team in this fourth session.

Overheads are likely to elicit some discussion (salaries are overheads).

Fifth: Brainstorming Ways to Cut Expenses. By now, you will probably have only 3 or 4 of your original 100 ideas left. Now, do a detailed gross profit analysis with actual figures. The information you need (accurate revenue and expense figures) will take time to gather. Assign those responsible to gather that information.

Until now the focus has been on the enterprises. Here the focus turns to routine costs. Every dollar you can cut from maintenance expense can be applied to wealth-generating expenses. This session should include everyone and could even benefit from the creativity of a few outsiders.

List your current expense headings and make sure everyone understands them. To be effective, brainstorming must be done quickly. The resulting list will require sifting, testing and often additional information. Be careful that expense cutting doesn't cut corners and sacrifice quality.

Creating your Plan

Each income or expense item and the months the money will be received or spent for each enterprise are recorded on simple worksheets. There should be one worksheet for income and one for expenses.

The rate of consumption for any supplies purchased in bulk should be planed to avoid the surprise of having to order ahead of plan.

Each overhead expense category usually has its own worksheet—i.e. office expenses, salaries.

Livestock enterprises must track animals that breed, die, change age classification, are culled or have their wool or hair sheered within the year.

When the figures are finalized, they are transferred to a general spreadsheet with columns that correspond to each supporting worksheet. The main planning team is responsible for finalizing the figures and completing the spreadsheet.

Creating your plan involves four basic steps: plan the income; plan the profit; plan the expenses; and assess the plan before implementing it. *Each is discussed below*.

Plan the Income: Although you estimated income earlier when you determined gross profit for each enterprise, the figures may have changed by the end of the preliminary planning phase. For example, any action taken to address a weak link will mostly likely increase income estimates.

Plan the Profit: Now determine how much of total planned income to set aside for profit. The amount remaining will have to cover all expenses. Having made a preliminary allocation, you should already know the approximate total that you need for that.

Deciding how much to set aside as profit is subjective. The purpose is to place a ceiling on expenses. If you are deeply in debt and a large part of your income must go to debt service, a smaller percentage may be all you can afford to set aside. *Sometimes it is better psychology to first subtract annual debt servicing payments then determine how much of the remaining to set aside for profit.*

Plan the Expenses: If you have to reduce expenses to stay within the limits, start with maintenance expenses. If you can't trim enough off of those, your planned profit is probably too high. But, before decreasing it consider reducing some of the wealth-generating expenses—being aware that you will lose the additional income they would have generated. In the end you may have to accept a lower profit. *The decision to forgo profit in favor of investing in the business is fine as long as you are doing it intentionally.*

Now look at maintenance expenses again to see if you can shave them a little closer. If so, apply what you gain to those wealth-generating expenses that need every dollar they can get.

Asses the Plan before Implementing It: Once the expenses are finalized record their totals on the main spreadsheet. Check to see that the total income projected for each month offsets total monthly expenses.

Have your accountant assess the plan's tax consequences, calculate depreciation based on the IRS' tables rather than any figures you may have used and account for any "non-cash" income.

Monitoring the Plan: Modify the plan as you progress through the year. Events rarely transpire exactly as planned which is why you need to plan. *If you knew what was going to happen in the future, there would be no need for a plan.*

Financial Planning Basics

Reconciling Financial Decisions with Your Strategic Goal: Your goal starts with the *quality of life* you want, states the *forms of production* that will help you get it, and then describes the *future resource base* needed to sustain it. Also, a future landscape description is a critical part of your goal. Clarity in your goal allows you to avoid temptations that may show tremendous promise but that lead you in the wrong direction.

The Testing Guidelines: Reducing Decision-Making Stress.

The Testing Guidelines are: Cause and Effect, Weak Link, Marginal Reaction, Gross Profit Analysis, Energy/Money Source and Use, Sustainability, and Society and Culture. As you allocate resources in your financial planning process, your policies and projects must be reviewed through these guidelines.

Speed is essential when testing to keep from losing sight of the whole. But, both the Cause and Effect and the Gross Profit Analysis require a lot of thought and some calculations so do them first then pass quickly through the other tests. If you are dealing with a problem, focus on cause and effect. If you are comparing enterprises, focus on gross profit analysis. In deciding where to allocate money you are actually making most of the major decisions for the year.

The testing guidelines overlap which allows you to pick up in one what you might have missed in another. It may be very difficult to figure out where one or another applies but do not let that bother you—they all function together. *Just remember why you are doing what you are doing—making decisions that will guide you toward your holistic goal.*

Financial Weak Link: Generating Wealth

At any given moment there is only one weakest link that must be dealt with before moving on to any other link. Identify the year's weak link in each enterprise. Then, when allocating money for expenses, actions that address the weak link in an enterprise have priority. Such expenditures are *wealth generating* because they boost production and profit. Other expenditures usually maintain current levels of production.

The chain of production has three links:

1. Resource Conversion: Ranchers convert sunlight energy into a consumable product. Ideally, the money you invest in a weak link should be "solar dollars." The weak link in each enterprise shifts with time but your emphasis should be on the resource conversion link because you can probably produce a better range that will turn more sunlight into saleable products.

Each enterprise will always have a weak link at any given point in time. When you identify a weak link, you must implement an action to strengthen it. For example: buying feed does not strengthen the resource conversion link but building fences and buying land can.

2. Product Conversion: If you have too few animals to use the forage produced, product conversion is the weak link that can be addressed by increasing animal numbers.

3. Marketing (Money Conversion): There may be a gray area between the product and marketing conversion links. For example: if the poor quality of your wool is due to the quality of your animals, you have a product conversion weak link. However, if it is due to the wool being delivered dirty, it is a marketing conversion weak link. But *such a fine line distinction is not what is important. What really matters is that you recognize the problem and address it.*

Energy/Money Source and Use: Investing Soundly

Money derived from the mineral wealth of the earth is what Savory calls *mineral dollars.* And, money derived from plants grown by the sun is what he calls *solar dollars. Paper dollars* are the third form of money. These are based on human creativity and financial transactions. The basis for paper dollars is no deeper than the public's confidence in the government. Consequently they can be created and/or destroyed and are, therefore, very unstable. Rely on paper dollars at your peril.

You will be much better off if you measure your success in solar dollars. The more you rely on solar dollars, the more you are insulated from swings in prices and interest rates. This does not mean that you should ignore prices and interest rates but it does mean that you should take a nimble and flexible attitude toward them.

(Editor's Note to couch this more in economic terms: Nowadays and worldwide, fiat currencies printed by central banks are not backed by any commodity. Central bankers define inflation as a "general price increase" but that is (intentionally) misleading. Inflation is simply an arbitrary increase in the supply of fiat money and could not occur if the currency was commodity backed.

So the point here is to turn your "funny money" into hard, wealth generating assets (land, livestock, fences, water sources, etc) that provide immunity to the central bank's unpredictable manipulations. (Of course one has no other option but to keep at least enough funny money on hand to conduct day to day business and stay out of debt.)

Marginal Reaction: Getting the Biggest Bang for Your Buck

This guideline is, to some extent, the accountant's equivalent of the weak link test. But, it applies in non-quantifiable situations as well. Assume that more than one action will strengthen a weak link. Their providing the same return is highly unlikely when comparing their marginal reaction per dollar or hour of effort. For example: Suppose resource conversion is your weak link and there are several ways to address it: buy or lease land, develop water, reseed, hire herders, build fence, etc. The question is: which has the greatest marginal reaction?

Gross Profit Analysis: Bringing in the Most Money for the Least Additional Cost

Gross profit analysis allows you to compare enterprises or combinations of enterprises.

Living expenses, debt payments, full-time labor and many other costs will continue to exist no matter what you produce so they need not be considered when considering a new enterprise.

Only looking at the "bottom line" (net profit or loss), does not tell you much about the aspects of an enterprise that contributes the most toward covering your fixed costs (overheads).

Accountants often want to apportion fixed costs among enterprises but that only obscures the real contribution the enterprise is making toward covering total fixed costs.

Any expense that you cannot avoid during the planning period, regardless of your plan, is fixed. Any expense that arises as a result of your plan is variable.

Editor's Note: I prefer to refer to enterprise gross profit [revenue minus direct (variable) costs] as "contribution margin" because it reveals just how much the enterprise is contributing to covering overheads (fixed costs).

Key ways that gross profit analysis influences your planning:

1. It allows you to be more responsive to changes in the markets—sheep vs. cattle, yearlings vs. pasture cattle, etc.

2. You will make better and smarter use of the resources you already have (land, equipment, labor) vs. making "cost effective" new investments.

3. It makes you acutely aware of fixed costs. Thus, you can plan toward cutting those costs and getting rid of inefficient assets and thereby cut your exposure to paper dollars and compound interest.

To compare livestock options, you must reduce them to a comparable base--$/head/year, $/acre/year, $/standard animal unit/year, etc.

For example: Say you run sheep, commercial cattle and registered cattle and have a gross profit of, say, $341.10/sheep; $556.18/commercial cow and a loss of $353.40/registered cow. Now, if we use the standard ratio of 5 sheep to one cow, then the sheep yield ($341.10 X 5) or $1,705.95 per animal unit—drastically greater than commercial cows and even more so for registered cows.

Three purposes for gross profit analysis:

1. To compare enterprises.
2. To see how different conditions (price changes) might affect the enterprise.
3. To weigh the effect of diverting assets from one enterprise to another.

Comparing Scenarios: Gross profit analysis can be used to assess the impact of future events, for example, poor, average, and good prices. The same approach works for changes in supplements, calf crops, average daily gains, etc.

Comparing the Use of Assets: Gross profit analysis allows you to explore the effect of changing an asset from one use to another—i.e. would that piece of bottom land be best put into alfalfa or used as pasture for stockers.

Even if the two options you are comparing have similar bottom lines, the process will force you to find answers for such questions as: Which is the highest risk? Would owning machinery make the hay option cheaper? Should you raise your own stockers or buy them? And so forth.

Gross Profit per Unit: $/head, $/acre, etc. are written like fractions because they are. The top is divided by the bottom. For example: if 100 cows yield a gross profit of $24,000, then $24,000/100 = $240/cow. If the herd uses 800 acres, then $24,000/800 = $30/acre. If the herd would cost $50,000 to replace, then Gross profit/ $ invested = $24,000/$50,000 = $0.48 (commonly referred to as 48 cents on the dollar).

The Danger of Gross Profit Analysis: Gross profit tells you what each enterprise contributes toward covering fixed costs. However, only the final financial plan that combines all enterprises tells you whether they will cover all the fixed costs. Like any other technical gimmick, gross profit analysis should not replace common sense.

Editor's Note: Land & Livestock International uses (and teaches) a technique called "sensitivity analysis" which involves setting up Excel spreadsheets to conduct enterprise gross profit analysis. A segment of the spread sheet is designated for operator input of assumptions of the variables. This allows the manager to manipulate the assumptions and see instantly their impact on the outcome. It also allows the comparison of several different enterprises on the same spreadsheet.

Paraphrasing the "holistic" literature: Diversity of enterprises is a hedge against changing market conditions. By the same token, biological diversity is essential to sustainable ranching. For both reasons, it can be dangerous to rely on gross profit analysis alone.

Editor's Note: Sensitivity analysis as (explained just above) mitigates this problem.

Financial Planning Forms

(Editor's Note: Most of the forms referred to in this lesson are available at Holistic Management International's (HMI) web site. Most of them can easily be fabricated from an old fashioned accountant's spreadsheet or a more modern Excel spreadsheet or Word table. Furthermore, there are many computerized accounting programs (Simply Accounting, Quick Books, etc) that have features for creating budgets, calculating variances between actual and budgeted amounts and presenting them as an integral part of the financial statements.)

Before you can generate a single grand plan, you will need to work out the details—hours of labor, quantities of supplies, etc.

You will have four different forms to develop and monitor your plan. Most of them you can fabricate for yourself. However, understanding their use will help you understand the process.

1. ***Worksheets:*** to record the detail of each income and expense.

2. ***Annual Income and Expense Plan:*** To summarize all the worksheets. Results in a master plan and is the same form you will use to monitor the implementation.

3. ***Livestock Production Worksheet:*** Details livestock breeding operations. (This and the grazing planning sheet are probably the only two forms that would be worthwhile to order from HMI.)

4. ***Control Sheets:*** For reporting items that run counter to plan and record corrective actions and their responsible parties.

Standard Worksheet: Has columns for the months of the year and rows for the different expenses, income, decisions, etc.

The whole operation may take dozens of worksheets. Use only the final drafts to assemble the final plan. These can also be used for work schedules, inventory monitoring and any manner of monitoring and control.

Examples of Other Worksheets:

Being able to look at a glance at the biological year of a commercial cattle herd:

- Helps design culling policy or making changes in herd size
- Shows animals in each class for calculating supplement quantities, forage needs, etc.
- Helps plan sales policy—holding late calves, breeding back open heifers, etc.

Biological Year of a Cow Herd

WORKSHEET

Date: 2006
Planning Sheet Column Reference: _____
Worksheet No. 3

	Jan	Feb	March	April	May	June	July	August	Sept	Oct	Nov	Dec
Seasonal Year	←Calving→			←Bulling→					Weaning			
Mature Cows	←3+ years of age→			←Bulling→								
Heifers 2	←36 mo Calving→			←2nd Bulling→ 27 mo			30 mo	31 mo	32 mo	33 mo	34 mo	35 mo
Heifers 1	←24 mo Calving→			←1st Bulling→ 15 mo			18 mo	19 mo	20 mo	21 mo	22 mo	23 mo
Heifers	12 mo	13 mo	14 mo							9 mo	10 mo	11 mo
Steers	12 mo	13 mo and sold somewhere from here to 18 mo depending on market								9 mo	10 mo	11 mo
Culls										For sale anywhere from Oct on		
Bulls					—HAVE BULLS ALL YEAR LONG—							
Months of age Transfers												
Total												

150

Alfalfa Production and Sales Plan

WORKSHEET

Date: 2006 Planning Sheet Column Reference: ___ Worksheet No. 5

	Jan	Feb	March	April	May	June	July	August	Sept	Oct	Nov	Dec	
160 Acres Alfalfa										960 tons kept use on the ranch			
Harvest av. 3 ton/acre						480 tons	480 tons	480 tons				1,440 tons	
Sales		sell 120 tons				sell 120 tons	sell 120 tons					sell 120 tons	
Estimate Price		$100/ton				$80/ton	$80/ton					$90/ton	
Cash Income		$12,000				$36,400	$19,200					$10,800	$80,400
Total													

Fuel Use and Purchase Projections

WORKSHEET

Date: 2006 Planning Sheet Column Reference: Fuel Worksheet No. 14

Seasonal Year	Jan	Feb	March	April	May	June	July	August	Sept	Oct	Nov	Dec	Item Totals
Seasonal Year	← Feeding and Calving →				← Bulling →				Weaning	Sales	← Feeding →		
Manager's Suburban	200m	250m	250m	250m	250m	250m	250m	250m	250m	400m	100m	100m	2,800 miles
Suburban Off Ranch			4,000m					2,000m					6,000 miles
Jim Jones Pickup	200m	200m	150m	50m	50m	50m	50m	50m	50m	50m	200m	200m	1,300 miles
Jones Motorbike			100m	100m	300m	300m	300m	300m	300m	300m			2,000 miles
Jake Moncho Pickup	350m	350m	250m	250m	60m	60m	60m	60m	60m	60m	350m	350m	2,260 miles
Moncho Motorbike			150m	150m	400m	400m	400m	400m	400m	400m			2,700 miles
John Smith Mechanic P.U.	100m	100m	100m	300m	300m	300m	300m	500m	500m	500m	100m	100m	3,200 miles
Pickups miles/ gals @ 12 mpg	850m/71g	900m/75g	750m/63g	850m/71g	660m/55g	660m/55g	660m/55g	860m/72g	860m/72g	1010m/84g	750m/63g	750m/63g	9,560m/799g
Purchases off ranch	4,000 miles estimated		$866			2,000 miles estimated		$434					$1,300
Motorbikes miles/ gals @ 90 mpg			250m/3g	250m/3g	700m/8g	700m/8g	700m/8g	700m/8g	700m/8g	700m/8g			4,700m/52 g
Monthly gas consumption	71g	75g	66g	74g	63g	63g	63g	80g	80g	92g	63g	63g	To plan sheet by month
500 gals bulk purchase gas		$1,200						$1,200 approximately 147 gallons on hand at year's end					$2,400
Monthly Totals	$1,200	0	$866	0	0	0	$1,200	$434	0	0	0	0	$3,700

Holistic Management International

RANCH: River Bench
ENTERPRISE: Cattle
REMARKS: One year analysis

LIVESTOCK PRODU[CTION]

January 2006   YEAR OR MONTHS: April/May

A CLASS OF STOCK (Cows, Heifers, Calves, Bulls, etc.)	B BIRTH % EST.	C OPEN NO.	D AGE	E BIRTHS	F MONTH	G BUY	H MONTH	CLASS TRANSFERS IN	CLASS TRANSFERS OUT	I DEATH	J %	K SALE	L MONTH	M CLOSE AND OPEN NO.	N AGE	E BIRTHS	F MONTH	G BUY
1. Bulls		20				5	2		1	0				25				
2. Cows	90%	500						200	2					700				
3. Heifers - 2	70%	200						225	3	200	0			225				
4. Heifers - 1	90%	225						240	4	225				240				
5. Heifers		240							5	240	0			0				
6. F. Calves		0		396	2				6		0			396	4			
7. M. Calves		0		396	2				7		0			396	4			
8.									8									
9.									9									
10.									10									
11.									11									
12.									12									
13. TOTAL HEAD		1185		5										1982				

NOTE: RECORD YOUR ESTIMATES OF PERCENTAGE ACTUAL BIRTHS FOR THE VARIOUS AGE CLASSES OF BRED FEMALES IN THEIR ROWS IN C[OLUMN...]

14. CLASS OF STOCK	Bulls		Cows		Heif-2		Heif-1		Heifers		M. Calves		Bulls	
15. NUMBER SOLD / MONTH OF SALE	5	10	138	11	66	11	24	11	38	11	380	11		
16.														
17. AVERAGE LIVE WEIGHT	1500		950		900		800		450		550			
18. MEAT PRICE PER LB.	0.6		0.7		0.7		0.7		0.75		0.75			
19. INCOME PER ANIMAL	900		665		630		560		338		413			
20. WOOL/HAIR WEIGHT/MONTH OF SALE														
21. WOOL/HAIR PRICE/MONTH OF SALE														
22. WOOL/HAIR INCOME														
23. PLANNED GROSS INCOME	4500	10	91770	11	41580	11	13440	11	12825	11	156750	11		
24. NUMBER PLANNED TO BUY													5	
25. ESTIMATED PRICE/ANIMAL													500	
26. TOTAL COST AND MONTH													2500	12

ꝺUCTION WORKSHEET

DATE OF PLAN: June 2005

BUY	H MONTH	CLASS TRANSFERS IN	CLASS TRANSFERS OUT	DEATH	J %	K SALE	L MONTH	M CLOSE AND OPEN NO.	N AGE	E BIRTHS	F M	G BUY	H MONTH	CLASS TRANSFERS IN	CLASS TRANSFERS OUT	DEATH	J %	K SALE	L MONTH	M CLOSE NO.	N AGE
		1		0				25						1				5	10	20	
		2		14	2			686						2				138	11	548	
		3		5	2			220						3				66	11	154	
		4		5	2			235						4				24	11	211	
		5		0				0					380	5				38	11	342	
		6		16	4			380	8					6	380					0	
		7		16	4			380	8					7				380	11	0	
		8												8							
		9												9							
		10												10							
		11												11							
		12												12							
				56				1926										651		1275	

YEAR OR MONTHS: Aug/Sept (left) | YEAR OR MONTHS: December (right)

IN COLUMN (B) ABOVE.

ANALYSIS OF PLANNED SALES AND PURCHASES

Copyright © 2006 Holistic Management International

The following worksheet lets you know when your expenses and income will occur and predicts the effect of breeding and culling policies over a long period—e.g. you can tell immediately the effect of a 5% decrease in conception rate will have on herd size in three years.

Holistic Management International

RANCH: River Bench
ENTERPRISE: Cattle
REMARKS: 3 year plot to see rate of herd growth

LIVESTOCK PRODU(CTION)

YEAR OR MONTHS: 2006

A CLASS OF STOCK (Cows, Heifers, Calves, Bulls, etc.)	B BIRTH % EST.	C OPEN NO.	D AGE	E BIRTHS	F MONTH	G BUY	H MONTH	CLASS TRANSFERS IN		CLASS TRANSFERS OUT	I DEATH	J %	K SALE	L MONTH	M CLOSE AND OPEN NO.	N AGE	E BIRTHS	F MONTH	G BUY
1. Bulls		20				5	2		1		0		5	10	20				7
2. Cows	90%	500						200	2		14	2	138	11	548				
3. Heifers - 2	70%	200						225	3	200	5	2	66	11	154	36			
4. Heifers - 1	90%	225						240	4	225	5	2	24	11	211	24			
5. Heifers		240						380	5	240	0		38	11	342	12			
6. F. Calves		0		396	2				6	380	16	4			0		395	2	
7. M. Calves		0		396	2				7		16	4	380	11	0		395	2	
8.									8										
9.									9										
10.									10										
11.									11										
12.									12										
13. TOTAL HEAD		1185		792		5					56		651		1275		790		

NOTE: RECORD YOUR ESTIMATES OF PERCENTAGE ACTUAL BIRTHS FOR THE VARIOUS AGE CLASSES OF BRED FEMALES IN THEIR ROWS IN CO(LUMN)

14. CLASS OF STOCK							
15. NUMBER SOLD/MONTH OF SALE							
16.							
17. AVERAGE LIVE WEIGHT							
18. MEAT PRICE PER LB.							
19. INCOME PER ANIMAL							
20. WOOL/HAIR WEIGHT/MONTH OF SALE							
21. WOOL/HAIR PRICE/MONTH OF SALE							
22. WOOL/HAIR INCOME							
23. PLANNED GROSS INCOME							
24. NUMBER PLANNED TO BUY							
25. ESTIMATED PRICE/ANIMAL							
26. TOTAL COST AND MONTH							

...DUCTION WORKSHEET

DATE OF PLAN June 2005

		YEAR OR MONTHS 2007										YEAR OR MONTHS 2008						
G BUY	H MONTH	CLASS TRANSFERS IN	CLASS TRANSFERS OUT	I DEATH	J %	K SALE	L MONTH	M CLOSE AND OPEN NO.	N AGE	E BIRTHS	F M	G BUY	H MONTH	CLASS TRANSFERS IN	CLASS TRANSFERS OUT	I DEATH	J %	K SALE	L MONTH	M CLOSE NO.	N AGE
7	2		1	0		5	10	22				7	2		1	0		5	10	24	
		152	2	14	2	109	11	577						142	2	15	2	114	11	590	
		207	3 152	5	2	62	11	142	36					295	3 142	6	2	88	11	201	
		335	4 207	7	2	34	11	298	24					334	4 295	7	2	33	11	297	
		379	5 335	8	2	38	11	340	12					429	5 334	9	2	43	11	383	
			6 379	16	4			0		447	2				6 429	18	4			0	
			7	16	4	379	11	0		447	2			7	7	18	4	429	11	0	
			8												8						
			9												9						
			10												10						
			11												11						
			12												12						
7				66		627		1379		894		7				73		712		1495	

IN COLUMN (B) ABOVE.

ANALYSIS OF PLANNED SALES AND PURCHASES   Herd increase over 3 years = 310

Copyright © 2006 Holistic Management International

Each Livestock Production Worksheet supports two columns on the master plan: 1) income (livestock sales) and 2) expenses (livestock purchases). Other livestock expenses (feed, vet medicine, etc) are planned separately on standard worksheets.

Follow the steps below while referring to the worksheet above:

1. Use a separate worksheet for each enterprise—commercial cattle, sheep, etc.

2. For each, draw up a biological year worksheet for a clear picture of when animals will change age, class, etc. It is best to start with the youngest animal class and follow it through its life.

3. Column A: Record the different classes of livestock in the same order that they appear on your biological year worksheet—bulls, cows, heifers (bred), heifers (open), calves, etc.

4. Decide whether you are using the worksheet for one year or more than one year.

You can use multiple spreadsheets to track long-term livestock production
Each form has 3 similar sections.
> One can be used for one year and the others left blank, or
> The three sections can be used for one year divided into three, 4 month periods, or
> One worksheet can be used for three years

5. Referring to the biological year worksheet.
   Put the number of animals in each class in Colum C
   Age in months goes in column D
   For mature animals you can use 36+
   For young animals use the average age

6. Birth percentages for all bred females go in column B opposite the appropriate female in column A.

7. Mortality percentage for each class of stock goes in Column J.
   If you are doing three years, do this for every year.
   If you are only doing one year, use column J in the middle year as though they all occurred at midyear.

8. Using the percentage birth estimates (Col B), calculate births for each female group (check the biological worksheet). Group animals born and divide by 2 to estimate male and female calves. Record in Col. E.

9. List anticipated livestock purchases in Col. G with planned month of purchase in Col. H on the appropriate (class of livestock) row.

10. Follow each class of livestock through the year (easiest to start with the youngest). Note age, add purchases and calculate deaths (record deaths in Col. I) Enter those you plan to sell in Col. K and the planned month of sale in Col L.
    For changes in age class, enter the number OUT and the number IN (OUTS must always equal INS) in the shaded class transfer columns. (If you have already taken off deaths, do not do it again in the class you transferred them to.)
    Follow each group through the year and record the close in Col M with average age in Col. N. This becomes the opening column for the next year. (If you are using one worksheet for a year, close each 4 month period this way.)

11. Check your work. Add the livestock you started the year with and enter in row 13 Col. C. To that add columns E and G then subtract columns I and K and the answer should be the total you have under Col M. If not, you have made a mistake.

12. Analyze sales: Record each type of animal to be sold on Row 14. Look at Col K and transfer the number for sale in each class to row 15 with the month of sale from Col. L. Note that you have a spare row (16) in case you have more than one class of animal to be sold. For each class:
    Estimate weights in row 17
    Price per pound in row 18
    Income per animal in row 19
    Wool, hair weights and prices (if any) to in rows 20-22
    Calculate estimated gross income and record in row 23
    Transfer these total planned income figures to the main spread sheet.

13. Analyze purchase:

Record the classes of livestock to be bought in row 24.
Price per animal – row 25
Total cost – row 26
To the right of that—month of purchase from Col H above.
Transfer these final planned livestock purchases figures to the appropriate enterprise columns of main spreadsheet.

Step by Step Guidelines for Creating Your Plan

What follows is a detailed example of what was presented above. It is a procedure for creating a total ranch plan that accounts for both money and biological capital.

Planning the Planning: Laying the Foundation: Standard cost accounting procedures treat capital expenditures as "wealth generating" if they return enough revenue to pay for themselves plus an acceptable margin. This can lead to failure because wealth-generating potential also includes noncash items like soil productivity, water quality and training. *(Editor's Note: It does not necessarily lead to failure. All that is needed is a healthy awareness of the wealth generating potential of soil, water and training that can stem from the use of common sense.)*

You will find that Excel (or any sophisticated spreadsheet program) will be very useful and will save you a lot of time (and money) over the long run.

Creating Ownership: Who Should Be Involved? Select a team as inclusive as possible without being unwieldy.

- People involved at the beginning tend to care about projects through the end.
- The people responsible should be allowed to come up with the figures under their control
- When hard choices must be made, morale will be better if everyone has a chance to work on the solution.
- It takes a lot of work that might keep one person tied up year around.

Overview and General Scheduling: Routinely start the process two or three months before the beginning of the fiscal year. Most of the time will likely be spent on the first of the following three phases.

1. *Preliminary planning.* Review the current year's plan and gather information for the new plan.
2. *Putting the plan on paper*—if you have done a good job of researching costs, etc during the preliminary planning phase, you won't need but a day or two.
3. *Monitoring the plan*—spend a few hours once per month to monitor continuously through the year. Events rarely turn out as planned. If actual deviates from planned, act immediately to get back on track.

Preliminary Planning: Taking a Hard Look: In this phase, you will make decisions that bear directly on the income and expense figures you will enter onto the worksheets. There are *four basic questions* that must be answered:

1. Is There a Logjam, and If So, How Will You Address It? Ask yourself whether there is anything blocking you from making progress. Step back from the day-to-day operations to get a better perspective—try to identify that one "log" that needs to be removed to break the jam. This is not always easy, tidy or obvious. Sometimes it takes an outsider to see the problem.

When folks are new to this type of planning, the most common logjam is failing to change how decisions are made. Other common ones include:

- Government regulations (BLM, USFS) that prevent adequate animal impact
- Contractual agreements that have you locked into a particular type(s) of production.
- Lack of adequate water
- Lack of knowledge in handling large herds
- Location relative to specialty markets
- Ranch too small to be ecologically manageable—need to combine with others
- Lack of capital
- Properties too large to be effectively managed by available or willing people (common in Australia)

If overcoming a logjam involves spending money, this is the most important thing to spend money on.

Create a worksheet labeled: "Wealth Generating Expenses" and list expenses needed to address a logjam and note "logjam" beside them. If you have actual figures, include them. If you don't, research them.

2. Are Other Factors Adversely Affecting the Business as a Whole? Think broadly to identify things that are reducing efficiency and productivity and, if not addressed, might become a logjam at some point. In addition to the items listed above, common things include:

- No time for vacations or family visits
- Lack of computer skills
- Lack of family communication and cohesiveness
- Lack of communication infrastructure—internet access, e-mail, etc.
- Lack of expertise or connections for marketing value-added products

The remedies may or may not involve expenditures. For those that do:

Create a worksheet labeled "Adverse Factors" and list the items. If you have figures, record them. If not, research them. If any of them could potentially prevent you from addressing a weak link or are greatly reducing income, consider them "wealth generating" expenses and record them on your Wealth Generating Expense worksheet.

3. Are Current Enterprises "Profitable" and Have You Spread Your Risk? If gross profit for any enterprise is not as good as planned, determine why and make adjustments or drop it.

If your cull ewes sold for more than you planned, will that likely happen again?

You want to spread your risks. It is unwise to commit all your resources to one product next year because it was exceptionally good this year.

Use gross profit analyses to determine which enterprise(s) to allocate more funds or to drop if it is not providing a positive contribution to overheads—unless you have a special reason to keep it.

Create a worksheet labeled "Gross Profit Analysis." List each enterprise and record their gross profits. Note any modifications you intend to make and list any new enterprises you might want to add and their gross profit figures. With all the enterprises on the same worksheet, you can see how profitable you are likely to be next year. If the total gross profits are not enough to cover fixed costs, you are in trouble and need to modify existing enterprises or seek more lucrative ones.

Selecting Appropriate Enterprises: Remember:

- Managerial effectiveness is diluted by the number of enterprises one manage is responsible for. Being stretched too thin destabilizes the business as a whole. This might be overcome by contracting the management of the new enterprise to someone else.
- There is a direct relationship between management effectiveness and the distance to what is being managed. The more frequent the contact with the enterprise, the greater the chances of spotting trouble early.
- It is often easier to alter an existing product, or develop new uses for it, than to create something entirely new. There may be ways to "add value" to a crop with little additional effort.
- Possibilities may exist to collaborate or partner with other related businesses or organizations that product complementary products or services. For example, you might form a cooperative to exploit a niche market.

4. What Is the Weak Link in Each Enterprise and How Will You Address It? Give thought to the weak link in each enterprise—Resource (energy) conversion; Product conversion; Marketing conversion—and determine what is needed to address it.

Some or all enterprises may have the same weak link.

What starts out as the weak link in an enterprise might be addressed quickly and cheaply and another link then becomes the weakest.

Any expenses involved in eliminating a weak link are wealth-generating and given priority.

Create a worksheet labeled "Weak Link Expenses" and list all your enterprises. Write in the actions needed, determine the costs of each and record these on the worksheet.

Putting the Plan on Paper: Once you have identified the logjam, completed a gross profit analysis and identified the weak link, it is time to create the plan.

Things to Keep in Mind: Rancher excuses for failure are many but the main reasons are:

1. Allowing production costs to increase to optimistically estimated income
2. Borrowing against optimistically estimated income.
3. Allowing the promise of quick profit mask damaging side effects
4. Not spending enough time working out reality on paper or using methods that don't work.

To counteract these:

1. Calculate total gross income then set aside up to 50% as profit with the remainder being all you have for living and production expenses. You can allow less but if it is too small you are likely to exert too little effort in cutting expenses.

2. Warning: Sales people will push strategies for "higher yields, maximum gains, etc. instead of greater profits. Make your position clear as to your limits.
3. Keep your strategic goal in mind and test all actions and tools against it.

A worksheet showing how you arrived at your figures must back up all the numbers you enter on the spreadsheet—i.e. each column on the financial plan spreadsheet should be supported by at least one worksheet. These worksheets should be filed where you can find them for subsequent monitoring and control.

Step 1. Plan the Income and the Direct Expenses Associated with Each Enterprise. 1) Prepare one income and one expense worksheet for each enterprise. (For breeding operations refer to the Livestock Production Worksheet.); 2) Prepare inventory consumption worksheets for items you purchase in bulk that are directly associated with an enterprise.

Step 2. Plan Miscellaneous Income. Prepare a "Miscellaneous Income" worksheet for income sources not related to a specific enterprise.

Step 3. Categorize the Expenses. Think through the year and decide which of the following categories each expense falls under.

1. *Wealth-generating expenses* are those that generate new wealth by addressing a logjam and/or the weak link in an enterprise. You identified these during your preliminary planning and recorded them on the wealth generating and weak link worksheets.
2. *Inescapable expenses* are ones that you are legally or morally obligated to—e.g. debt service, land taxes, etc. List each amount by month due on an "Inescapables" worksheet.
3. *Maintenance expenses* are the remaining expenses associated with each enterprise and all fixed costs that are essential to maintaining present income but will not generate any additional. Most expenses will fall into this category.

Step 4. Plan All Other Expenses. This includes "fixed costs" not directly associated with any enterprise. Some can be grouped on a "Miscellaneous" worksheet while others will need separate worksheets labeled accordingly. For example, fuel will require a worksheet to estimate consumption, times of bulk buying, etc.

Step 5. Transfer Income Figures to the Spreadsheet. Set up an income column for each enterprise on the Annual Income and Expense spreadsheet. Also include one last column for miscellaneous income. Transfer the figures from each supporting worksheet to the appropriate columns and months. Total the income down each column and across each row. The sum of the totals down and the sum of the totals across must match or you have made an error.

Step 6. Allocate Funds to the Logjam (If there is one). The first column on the expense side is the most important. Label it "Logjam." Transfer numbers from the Wealth Generating Expense worksheet into the plan row and appropriate months.

Step 7. Plan the Profit. Do this before allocating money to any other expenses except for the logjam. This places a ceiling on your expenses. To proceed:

1. Reduce the gross total on the income spreadsheet by up to 50%. (For some, this would be demoralizing so 20 or 30% would be enough.)
2. If you have incurred a large debt, the money remaining might depress you. So, subtract the debt payment and then cut what is left.

3. Record this profit in the next column of the spreadsheet (or the first column if you have no logjam). Label the column "Profit" and enter the numbers in row under the months you plan to set it aside.

The following figure shows the set up for the "Annual Income and Expense Plan." This spreadsheet could very easily be adapted to an Excel spreadsheet. Also, most modern accounting programs (Quick Books, Simply Accounting, etc) have features that allow the user to set up budgets and display actual vs. budgeted and difference amounts on their financial statements.

Land & Livestock International, Inc. uses a cost accounting system that supplies this information.

Annual Income and Expense Plan

				Column headings for income or expenses			
January	PLAN						
	- ACTUAL						
	DIFFERENCE						
CUMULATIVE DIFFERENCE TO DATE							
February	PLAN						
	- ACTUAL		Planned amount for month (done in pencil when by hand)				
	DIFFERENCE						
CUMULATIVE DIFFERENCE TO DATE							
March	PLAN						
	- ACTUAL		Actual amount each month				
	DIFFERENCE						
CUMULATIVE DIFFERENCE TO DATE			Difference between what was planned and what actually happened				
April	PLAN						
	- ACTUAL						
	DIFFERENCE						
CUMULATIVE DIFFERENCE TO DATE			Total difference accumulated by this month from start of financial year				
May	PLAN						
	- ACTUAL						
	DIFFERENCE		Months listed to match the financial year for the farm				
CUMULATIVE DIFFERENCE TO DATE							
June	PLAN						
	- ACTUAL						
	DIFFERENCE						
CUMULATIVE DIFFERENCE TO DATE							
July	PLAN						
	- ACTUAL						
	DIFFERENCE						

Step 8. Transfer the Inescapable Expenses (If Any) to the Spreadsheet. Be sure that they are, in fact, unavoidable—i.e. by definition. For example, have you explored all the options for structuring debt – i.e. converting debt to equity or a sale lease-back arrangement? If so, allocate what you must. Transfer the numbers from the "Inescapables" worksheet's next column and label it "Inescapables."

Step 9. Transfer the Remaining Expenses to the Spreadsheet. Transfer the wealth-generating expenses first. Divide these into 2 groups: 1) those that must have 100% of the money needed or they won't happen and 2) those that could work with a partial allocation. Allocate all that is needed to the first type. Allocate a minimum to the second type while being aware that you may increase the amount before you are through planning. An example is fencing—if you allocated the maximum, you would probably run out of money before you finished the plan.

Transfer "adverse factor" wealth-generating expenses from the Wealth-Generating worksheet and label the column "Adverse Factors."

Transfer the "weak link" expenses into the appropriate enterprise columns labeled "WL" plus the enterprise name. Since you will have other expenses associated with each enterprise, count the number of worksheets you have for each enterprise and leave enough blank columns to handle these.

Transfer the numbers from the enterprise expense worksheets to the appropriately labeled columns.

Transfer monthly consumption numbers (volume, weight, etc) to their separate columns so you can monitor consumption over the year.

Transfer the fixed costs from their worksheets with each column having the same name as the worksheet and include columns for any inventory consumption worksheets.

Total the columns and rows and, again the totals must be the same. Enter this figure in the lower right hand corner of the spreadsheet.

Step 10. Bring the Plan into Balance. Does income cover expenses? If not, cut, paste, compromise and fit everything into one sound plan. Start with maintenance and cut any that are not absolutely essential or can be put off for the year without damage. Apply any cuts to wealth-generating expenses that received only minimum allocations but needed every dollar they could get. Place every dollar you can where it provides the highest marginal reaction. Consider reducing profit if the money will be put toward one or more wealth-generating expenses—i.e. invest the profit in growing the business.

Step 11. Determine Where to Invest Your Profit. When the plan is in balance, income will cover all expenses. Consider reinvesting profits in wealth-generating expenses or somewhere to spread your risk or maintain liquidity for emergencies. Also keep the tax consequences in mind.

Step 12. Check the Cash Flow. Just because the bottom line is positive doesn't mean that you will have the cash you need when you need it. Set up two columns on the far right of the spreadsheet, one headed "Monthly Surplus/Deficit" and the other "Bank Balance."

Calculate the monthly surplus/deficit for each month (total month income minus total month's expenses) and record the applicable row.

Calculate the bank balance by starting with whatever cash will be on hand when you begin the year in the block just above the first plan row. Add any surplus or subtract any deficit for the month to get the predicted bank balance.

Adjust the plan in any month that shows a negative bank balance by rescheduling purchases. If you cannot juggle the numbers enough to avoid deficits over several months, you will have to borrow enough to get you through the period.

Re calculate the totals down and across and make sure they are the same.

Editor's Note: The usefulness of an Excel spreadsheet in the above calculations is obvious.

Step 13. Figure Debt Costs in Relation to Cash Flow (If Applicable). If borrowing is involved, work out the full cost by calculating monthly interest. Do this analysis to the right of the expense columns.

Simple Overdraft: Use three columns: 1) Monthly Surplus/(Deficit); 2) Bank Balance; 3) Bank Interest Owed. Calculate each month's surplus or deficit separately by subtracting total monthly expenses from total monthly income and record in that month's "plan" row. Just below the "Bank Balance" heading, above the "plan" row, record the estimated cash on hand at the beginning of the year. Then add or subtract the monthly surplus or deficit and enter the result in the bank balance "plan" row. If it is an overdraft calculate the interest on it for that month and enter it on the "plan" row under "Bank Interest Owed." (See Example C, Holistic Management Handbook, pg 46)

For the following month, start with the overdraft from the previous month, add the interest to it than add or subtract the monthly surplus or deficit for the next month. Enter any expected overdraft on the "plan" row under "Bank Balance. (See Example D, Holistic Management Handbook, pg 47). This overdraft then becomes the basis for interest calculations for month 2. Assume that income exceeds expenses in month three to the extent that the bank balance becomes positive. No interest will be calculated.

Loans and Notes: Use another 3 columns to the right headed: 1) Loan Payments; 2) Loan Balance; and 3) Loan Interest Owed. Any adjustment down made in the principal must be reflected in the "Bank Balance" column.

Step 14: Assess the Plan. Two important Checks:

1. *Is the plan sound when projected forward?* To project the plan forward, transfer the figures to new worksheets then go through each item and make the changes you know will apply in the following year. (For example, if you plan to buy a new scale this year, then that will not be an expense that you will incur next year.) Now, look at the bottom line—the rate of loan payoff and the peak of indebtedness during the year. If those are OK, repeat the process for as many years as necessary to get a clear picture of the trend of the business.

2. *Is the plan sound from an overall business point of view?* Will the plan produce a profit or loss? Will it leave you in an acceptable position relative to your strategic goal? Remember: You may have allowed for depreciation but did not use the IRS tables. Nor have you accounted for increases or decreases in net worth—unsold produce on hand at the end of the year represents an increase in wealth. Calculate the difference between the beginning and ending values of assets including inventory. Add cash excess planned and deduct the allowable depreciation. Knowing the tax consequences will be useful so planning should be completed at least a month prior to the end of the tax year. If things do not look good, re-plan immediately until the results are satisfactory.

Monitor Your Plan to Ensure Profit: (See Figure 1.18, Holistic Management Handbook, pg 50)
Monitor at least monthly using these guidelines:

1. Establish a means to obtain the actual income, expense and inventory figures.
2. Enter the actual figures *in ink* on the second row ("Actual") of the spreadsheet.
3. Calculate the difference between planned and actual and enter on the "Difference" row. Enter adverse figures in red ink.

4. Record the accumulated difference in the 4th row again using red for adverse differences. This will alert you if small differences that are ignored accumulate into a serious drift away from the plan.

Consider all the major adverse deviations in detail by going back to the original worksheets. These may be due to inexperience at first but you will get better with time.

Income items that are seriously adverse can be controlled only by cutting expenses. If expenses are adverse, control the item itself. Never balance a surplus in one column against a deficit in another.

Lesson 9: Biological Monitoring

Introduction to the Basics of Biological Monitoring

What you want to happen should be incorporated into your long-term strategic goal. At that point, it becomes your responsibility to make it happen. Planning is necessary to that end.

However, only very rarely do events occur according to plan. Monitoring involves watching for deviations from the plan so that corrections can be made. So, the process becomes one of: plan-monitor-control-re-plan. In other words, the process is a continuous circular loop. If you do not respond to the feedback, you fail to complete the loop. Completing the feedback loop is what makes management proactive vs. reactive.

Land management practices can lead to unintended consequences. Therefore, always assume that you could be wrong and determine what criteria you need to monitor that will give you the earliest warning of adverse consequences. You are *not* trying to record change. Instead, you are trying to steer all changes in the direction of your long-term strategic goal.

In conventional resource management, it is common to monitor changes in plant and animal species. However, by the time such changes become observable, it is already too late. You do not need a mass of data with little practical significance. *Soil surface changes are important because they precede most changes in plant and animal populations. In fact, the earliest observable changes are most likely to be at the surface of the soil—plant spacing, litter, soil organic matter, insect activity, presence of seedlings, quality of runoff water, etc.*

Monitoring is pointless if you fail act to correct deviations from the plan. A prime example is a pilot project initiated by the uS Forest Service in Arizona. A goal was formed with animal impact and planned grazing as the primary tools. For an eight year period, the rancher operated according to plan and the Forest Service monitored the results. Each year the monitoring showed that soil capping was decreasing but soil erosion and plant spacing was increasing. A simple adjustment in animal numbers and behavior during the first couple of years would have fixed the problem but *no control was implemented*. So, at the end of eight years the project was declared a failure (and "proof" that the method did not work).

In sum, a plan serves little purpose unless its implementation is monitored and deviations are controlled. You must plan, monitor, control and re-plan because that is the only way you can make happen what you have said that you want to happen in your long term-strategic plan.

The land is complex. So, any time you plan *always assume you could be wrong.*

The livestock industry traditionally monitors animal performance (conception rates, daily gain, weaning weights, etc) but *yield per acre is more important to profit than yield per animal.*

You want the earliest possible warning so you can make changes before too much damage is done. So, as you apply the tools (technology, animal impact, fire, grazing, rest or living organisms) determine what criteria you can monitor that will give the earliest warnings of adverse change. *Monitoring changes in species is common but comes too late. You want to detect changes well before that happens.*

Monitoring must be addressed on three levels:

1. Cultivate a general awareness of the four ecosystem processes (water and mineral cycles, energy flow and community dynamics) and how the tools affect them.
2. Each year assess the soil surface based on one of the procedures described below.

3. If you are managing livestock, you must also monitor growth rates, water supplies, etc.

These three monitoring levels complement each other and quickly reveal the dynamism in all landscapes. If you have not looked at land this way before, no land will ever appear the same again.

General Observations

Historical Data: Your monitoring will be more beneficial if you know where you are starting in relation to the past. If you have read about what conditions existed a hundred years ago, you are better able to assess the land relative to its potential.

When you begin, photograph characteristic sites, take notes of their condition, and try to reconstruct what happened in the past.

Back up your general observations with fixed-point photographs and hard data from routine monitoring.

The Future Landscape: In forming your strategic goal, you described the four ecosystem processes as they will have to be functioning in the future. Progress toward these four processes is what you will be measuring.

After monitoring the first time, summarize the status of the four processes so you will have something to compare against subsequent years.

Photo Monitoring: Photographs taken of the same scene in different seasons across a period of years show changes better than any other record. You should carry a camera like a pair of fence pliers.

Digital cameras are best because you can check your photos immediately and reshoot if necessary. You can also store them on a computer—just be sure you set up a system that allows you to identify and retrieve the photos you need. Also, make sure you have a backup.

The Surface of the Soil is the Key: The earliest changes are most likely to occur at or near the soil surface. If all is going as planned, no adjustment is necessary. Otherwise, diagnose what went wrong and develop alternatives.

You are looking for basic information you can measure and understand, not a mass of data of little practical use.

Brittleness: Environments can be classified according to how well humidity is distributed over the year and how quickly dead vegetation breaks down. Environments at either end of the scale are likely to respond similarly to most of the management tools.

However, they respond differently to rest. *Total rest* is the withholding of any disturbance for an extended period of time. *Partial rest* occurs when grazing animals behave so calmly that a large portion of the soil surface remains undisturbed.

In brittle environments, either form will increase bare ground. In non-brittle environments either form leads to ground covered by vegetation. In the tropics (a high-rainfall but brittle environment due to seasonality of rain), rest will damage grassland and move secession toward woodland with enough leaf fall to provide ground cover.

The poorer the distribution of humidity (particularly in the growing season), the more brittle the area—even if total rainfall is high. Very brittle environments commonly have a long period of non-growth that can be very arid.

The Four Ecosystem Processes: Key Indicators

All four ecosystem processes need to function well to sustain the landscape described in your strategic goal.

Succession and Community Dynamics: Community complexity is the key indicator of stability, resilience, productivity and health. Identifying whether a community is increasing or decreasing in complexity is subjective but becomes easier if you think in terms of *succession*—bare ground to algae to lichens to moss then to grassland.

"Climax" is the end point of succession but few environments are stable long enough to determine what that really is. Many communities where large grazing animals have been absent for a long time have been defined as "climax" and actually used as a measure of range management success.

The level and direction of succession can be judged by:

- *Simplicity vs. complexity.* A wide diversity of species (vs. large numbers of a few species) indicates an advanced community. Monocultures indicate a lower level. Seasonal diversity is also important—i.e. in some areas an absence of either cool or warm season grasses suggests a simplified community.
- *Annual vs. perennial.* Dominance by annual plants indicates a low level of succession. Perennial plants contribute much more to soil cover and stability and also give a more consistent picture of community health.
- *Presence of certain plants and animals.* Most species thrive only at certain levels of succession. Rodents and harvester ants are low on the ladder of succession. Moose and fir trees are high.
- *Status of youngest age class.* The presence or absence of young plants is the earliest indicator of the direction a community is moving. *The absence of young perennial grasses foretells decline.* Young animals or plants that increase diversity or fill high-successional niches indicate advance.
- *Presence of woody species* reflect an advance in succession. In brittle environments, woody species often flourish in declining grassland. This is not an advance when it represents a loss of diversity. In such cases you may want to hold the community at the grassland level and not allow it to advance into woodland.
- *Status of ground cover.* Capped, bare soil is the bottom end of succession. Therefore, *an increase or decrease in space between perennial plants (especially grasses) is the earliest sign of change.* Changes in litter and capping often precede changes in succession. In most arid/semi-arid grassland, litter is the most important indicator of which way the community will move because it provides most of the soil cover.
- *Remnants of old communities* that once thrived document the land's decline, indicate its potential and provide seed for a comeback.
- *Economic uses.* When an area's livestock industry has to shift from cattle to sheep to goats, the community is generally declining.

Water Cycle

Erosion: Erosion (obvious gullies and less obvious sheet erosion) indicates a non-effective water cycle.

Here are some of the signs:

- *Pedestaling of plants and rocks.* Wind or water has carried bare soil away and left plants and rocks up on little platforms.
- *Flow patterns on bare ground.* Moving water or wind scours the soil, leaving patterns.
- *Litter banks.* Piles of litter jammed between plants and rock form small silt traps. They are a sign of more runoff than desirable but do stop some of it.
- *Siltation in low points.* Silt dropped behind check dams, on gentle slopes below hillsides or in streams.
- *Splash patterns.* Raindrops dislodge and break up large particles of soil when they hit bare ground. Note the height of the mud splatters on plants and fence posts.
- *Dunes* are a product of wind erosion and often form far from the erosion site.

Permeability and Effective Rain: Changes in soil permeability can cause changes in the water table and the disappearance or reappearance of springs. It can be estimated by pouring a quart of water on the ground and timing how long it takes to disappear. Wait 5 minutes and probe with a shovel to find out how effectively it has penetrated. A similar check after a light rain will give you an idea of how much growth response to expect from that rain.

Plant Habitat: Plant types will vary as soils range from dry to moist, from sealed to well aerated, from well drained to waterlogged and as water tables rise or fall. Poor aeration occurs in both waterlogged and capped soils. It affects plant growth as much as a lack of water because it decreases the effectiveness of the water cycle.

Mineral Cycle: This boils down to three questions you can answer by observation.

- Are minerals visibly cycling?
- If not, what happens to them?
- How well are corrective measures working?

The health of the cycle shows in the following:

- *Breakdown of litter (especially dung).* Poor cycles allow dung to linger for years. In brittle environments watch the amount of dead material that oxidizes and turns gray. In less brittle environments, look for distinct litter layers. In a well managed soil you should have a hard time determining where the litter stops and the soil begins.
- *Activity of soil organisms.* Most soil organisms are microbial but worms, ants and burrowing rodents enhance mineral cycles. With low biodiversity these can explode into annoying numbers. The remedy is increasing diversity. The offending species remaining will no longer dominate.
- *Presence of plants with varying root depth.* Deep rooted plants recapture and return leached minerals to the surface from deeper layers of the soil.
- *Livestock consumption of mineral supplements.* This often reflects missing elements in the natural mineral cycle. They may add enough of some trace minerals to the soil through their dung to reduce the need for supplements.
- *Deficiency symptoms in plants and animals.* Poor mineral cycles can produce rough coats, infertility and other problems in animals. Symptoms in plants include leaf yellowing and/or curling, developing brown spots, etc. Soil tests can tell some of the story. But, although the soil may contain the minerals, they may be unavailable to plants.

- *Soil pH, sodium and salination.* Acidity or alkalinity makes nutrients unavailable to plants. Water carries salts with it that are left at the surface by evaporation. Excessive sodium destroys soil structure and the problems are exacerbated by the lack of organic material.

For insight into the mineral cycle, monitor community dynamics with an eye toward:

- Plant diversity—particularly nitrogen fixing legumes;
- Agents of decomposition including animal impact;
- The trend (increasing or decreasing) of soil cover and organic material
- Varying root depths.

Energy Flow: Energy flow is a function of the total area of leaves actively converting sunlight into forage, the length of time conversion goes on and what happens to the forage afterward. Generally, high energy flow is characterized by:

- Abundance of broad-leafed plants that grow rapidly
- Close plant spacing
- Rapid growth unimpeded by soil problems
- Plants active through the longest possible growing period

Low energy flow is characterized by:

- Waxy or narrow-leafed plants with slow growth rates
- Gray, oxidizing grasses
- Wide plant spacing
- Slow plant growth due to soil problems

Soil Capping: The hard crust inhibits community development, blocks mineral cycling and increases runoff which causes erosion. Animal impact is the most practical way to break up the cap and incorporate organic material into the soil. Soils covered by litter and living vegetation will not cap.

Four degrees of capping:

- *Mature capping* results from long rest. Low seccessional lichen, moss and algal communities give it a dark tone. It sounds hollow when tapped. In brittle environments, it is often the last life left in a deteriorating environment.
- *Immature capping* has been broken in the past but is still hard enough to prevent water penetration and seedling establishment.
- *Recent capping* is a result of recent precipitation over a broken surface. If left unbroken it will eventually become mature.
- *Broken capping* is the result of recent animal impact. The soil is open to moisture and seed germination and establishment.

Reading Plant Forms: Plants forms often tell the history of management.

Overgrazed Plants: Learn to look for overgrazed plants instead of overgrazed ranges. Chronic overgrazing affects vast areas. But, some of the same symptoms occur when plants (and soils) are over-rested. Continuously be on the lookout for individual plants that have been overgrazed because of either too long a grazing period or too short a rest period.

- *Distorted growth:* Many plants respond to overgrazing by flattening out and forming a tight mat often referred to as *sod bound*. Other grass species hide new leaf behind spiky stalks formed by old stems.
- *Dead centers:* Overgrazed bunch grasses lose energy reserves in the stem bases or crowns, suffer root loss and die back at the centers. The old oxidized material normally remains until removed by fire or termites.
- *Disappearance:* Some plants simply disappear with remnants being found in areas that animals cannot access.

Over-browsed plants (Non-grasses—shrubs, trees, etc):

- *Distorted growth or hedging.* Branches make knobs at the end where new sprouts make a dense cluster (gardeners create hedges by repeated clipping). Leaves hide behind spines or old twigs or lie flat against the bark.
- *Browse lines.* Trees lose all foliage below the reach of animals. Some show knobs and bristles of over-browsing on lower branches but long plumes of growth above.
- *Disappearance.* Any species being over-browsed is unlikely to produce young offspring.

Over-rested Plants: Over-rest followed by overgrazing is the most frequent cause of degeneration of rangelands. The successional shift toward woody plants leads to widening plant spacing and loss of organic matter. Symptoms include:

- Old growth that remains into the subsequent growing season(s) and becomes gray or black.
- Plants that have not been grazed recently and have dead or weakened centers.
- Weakened root systems. Old growth is present and the plant can easily be pulled up by hand.

Identifying Species: Don't waste a lot of time learning the names of plants. Names are just the knob to the door leading to more useful knowledge. All organisms fill certain niches. The more you learn about how they fit together, the more you will understand land dynamics. More important than the name is learning what species need to thrive, how they reproduce, what preys on them, etc.

You can get a good start on your own limited habitat in a short time. From then on, teach yourself by observation, discussion with knowledgeable people, correspondence and the internet.

What Grazing Patterns Can Tell You:

Low Stock Density: If animals gnaw grass right down to the ground in paddocks that also contain untouched, rank plants, the problem is likely to be stock density. If livestock are spread too thin, untouched areas may be inedible by the time they return.

This phenomenon is called "patch" or "all or nothing" grazing and the signs are easy to see—sharply defined un-grazed patches with extreme grazing of other areas. Typical of low-density grazing, heavily grazed plants border patches of un-grazed plants. Old mature plants are not grazed. They oxidize while many other plants are severely overgrazed—so much so that soil is exposed and becomes capped making it more difficult for new plants to establish.

Habits and Routines: Livestock learn routines and they can train cowboys to follow them. For example, they discover that if they bawl loud and long enough at the gate to the next paddock, someone will assume they are starving and let them in. Giving in and reducing each grazing period by a day or two (e.g. faster moves) means overgrazing which means less production which raises pressure for faster moves in a downward spiral.

If livestock know that they will always move into the adjacent paddock, they will crowd that fence line. This often produces overgrazing along one side of the paddocks.

Trailing is also a habit. These routines can be broken by moving stock through different gates, relocating supplements and/or using temporary electric fence.

Look for signs of destructive routines and vary them in your planning.

Living Organisms (Community Dynamics): Monitoring living things to understand them in the context of community dynamics enables you to use living organisms as a tool to your advantage. More often than not, problems will not disappear unless you correct the distortion in the biological community that produced them.

Biological Monitoring Methods and Techniques: The following procedures emphasize soil surface conditions and plant density.

Almost all stockmen weigh their animals at least once a year. But, livestock are only the broker in the marketing of solar energy. So, it makes better sense to "weigh" the principle agent in the transaction—the land. Just as you wouldn't trust a stranger to weigh your stock, neither should you trust anyone but yourself to monitor your land.

Two Monitoring Procedures

- *Basic monitoring* involves taking pictures of fixed plots and making notes of what you observe. It shouldn't take more than a day per year.
- *Comprehensive monitoring* is more detailed and therefore is good for monitoring public lands which require detailed records. This may take a few days each year.

Basic monitoring does not produce "quantitative" data but comprehensive does. Neither of these techniques are substitutes for day to day monitoring of growth rates.

Your monitoring is unlikely to satisfy academics because their approach is from a totally different perspective. You are monitoring to bring about intended change, not to simply record what happens. The main value to you is that at least once a year you are compelled to closely observe the source of your livelihood.

Gathering the Data

Transects: The *basic monitoring* procedure *uses straight-line transects* with *five* fixed sampling points.

The *comprehensive* procedure *uses a fixed-area transect with up to a hundred random sampling points* located by throwing a dart backward over your shoulder.

Transacts are *sited in representative areas* with *three to five* being *adequate on uniform land*. You will need *more if you have many different ranges types* or varied terrain.

The Time to Monitor: You should monitor at *the same biological time each year*. Monitoring *during the most active part of the growing season yields the best information*.

If possible, have the *same person* collect the data each year. New people will inevitably change procedures slightly. Work with them to minimize the problem.

Monitoring Criteria May Change: Criteria may change as the land improves. If you have produced dense perennial grass, it becomes pointless to monitor spacing between plants and soil capping because there won't be any. You might turn your attention to increasing energy flow and, therefore, decide to monitor for an increase in broader-leafed plants.

Closing the Circle of Management: Monitoring is useless if you don't take action when you observe adverse changes. Thus, the final step of the process is to record the action you plan to take.

If you think through the tools you've used and how they affect the land, the action you need to take is usually obvious. If you remain uncertain about one or more actions, you might test them on a small piece of land or look for evidence elsewhere on your property—i.e. look for *positive* deviations with an eye to discovering why.

Basic Monitoring

Basic monitoring is simple, requires minimum work, allows you to closely observe changes and provides a good record in the form of digital photographs.

Equipment Needed

- Camera (preferably digital)
- 100 ft tape
- 1 yd^2 PVC frame
- Pad of paper (or dry erase board) with pencil or marker
- Clipboard with pen or pencil
- Monitoring data forms (5 per transect)

When establishing transects:

- Metal posts (2 per transect) driven into the ground at either end of the transect line.
- 5 short rebar stakes (or survey "whiskers" with spikes) per transect. (Survey whiskers are available at survey supply stores.)
- Heavy hammer

You will establish straight-line transects with 5 monitoring points each. You will stand over the square PVC frame and take photos straight down. It is sometimes hard to avoid casting a shadow on the plot. Orienting the transect line east-west might help. Make a stand to hold the camera directly over the plot at the correct height. The stand can be made of light angle iron or galvanized pipe threaded so the camera can be attached.

Establishing Transects: Carefully select monitoring sites that are either typical of the general area and/or where you want to produce a lot of change.

Hammer the first steel post into the ground being sure posts stick high enough above the ground to be visible.

Extend the 100 ft tape and hammer in the second post at the 100 ft mark.

Hammer in a short rebar rod (or survey "whiskers") at five equally spaced (10, 30, 50, 70 and 90 ft) points along the tape.

Recording Transect Information: You will need five Basic Biological Monitoring Data forms per transect. Assign a number to the transect and to each plot (e.g. 1-1, 1-2, etc). Note the date and the name of the recorder. Use the back of the form for other information you feel is important.

Taking Photos: Take *two general view photos, one from each end of the line.* Include 1/3 sky and 2/3 foreground. When you print these, label them with the transect number and direction (e.g. 1E would be the eastern view from transect one). Note the photo information at the top right hand corner of the first form.

Write the identifying detail (1-E, etc) and date on a large piece of paper or board and place it within the camera's field of view and close enough for the writing to be visible.

At each marker along the line, lay one corner of the PVC frame over the rebar peg and one side flush with the tape. Write the plot number and date on a piece of paper (or dry erase board) large enough the letters can be seen in the photo. Place it in one corner of the frame. Position the frame and paper/board exactly the same each time you retake the photo.

Photograph each of the five plots with the camera directly above the center of the PVC frame (be sure the frame is in view of the camera). A stand to hold the camera at a consistent height and in the proper position is highly desirable.

Recording Data: Note the photo number at the top of the monitoring form and record your observations (refer to the form on the following page):

- *Soil surface.* Bare, capped, broken, covered with liter, covered with algae and lichens, hard, soft, porous, signs of erosion (pedestaling, siltation, etc.).
- *Animal Sign.* Hoof or footprints, scratch marks, dung. Minute trails, burrowing, birds and reptile nests, feathers and burrows, insects sighted, etc.
- *Litter.* Describe quality and condition. Fresh, old, breaking down, etc.
- *Perennial grass condition.* Healthy, mature, young, seedlings, dead or dying, over-rested, overgrazed, etc.
- *Grass species.* List names, if known. If not and you believe it to be important, take a sample and have someone identify if for you.
- *Other plants.* Forbs (such as legumes), noxious plants, plants that provide feed at critical times. Are they annuals, bi-annuals or perennials?
- *Points of interest.* Include things that might not show well in the photo, things that interest you but are not covered above, a trend that you want to keep an eye on, etc.

Biological Monitoring Data - Basic
(Five needed per transect)

Property _____ Transect/Plot Number _____ Photo No's _____

Date _____ Examiner(s) _____

1. Soil Surface. Describe the nature nature of the bulk of the soil surface between plants. (Is it bare, capped, broken, covered with litter, covered with algae and lichen, hard, soft, porous, etc.? Are their signs of soil movement/erosion, such as pedestaling, siltation in low points, etc?)

2. Animal Sign. What signs of animal life are present (small or large animals, birds, insects, reptiles)?

3. Litter. If there is litter present, describe its quality/ condition (fresh, old, or breaking down so it is hard to distinguish where litter ends and soil begins).

4. Perennial Grass Condition. If perennial grasses present, describe their condition. (Are they healthy, mature, young, seedlings, dead/dying, overrested, overgrazed?)

5. Grass Species. List grass species in the plot if you know their names.

6. Other Plants. List or comment on other non-grass plant species present (legumes, forbs, etc.).

7. Points of Interest. Note any other points of interest, including things that might not show well in the photo.

Analysis

You monitor so you can make happen what you want to happen. Based on your observations, you can draw conclusions about where you are relative to your strategic goal and what you need to do to be sure you keep moving toward it.

Set Up the Monitoring Analysis Form: You need one Basic Bio. Monitoring Analysis form for each transect. Record the details at the top. (See the form on the following page.)

Record your Analysis: Review your data forms and summarize what you found relative to the questions asked on the form. Record your answers to the following questions:

1. *Future landscape description.* What are you trying to achieve expressed in terms of the four ecosystem processes? Describe those specific to the area surrounding each transect. Also, indicate whether you are trying to create specific landscape features (brushy areas for example). Your answer to this question will likely be the same every year.
2. *Progress check.* What progress have you made this year relative to last? Review each data form and note specific changes in terms of community dynamics, water and mineral cycles and energy flow. For example, if the soil is bare with mature capping what does that tell you about the water cycle?
3. *Influencing factors.* Whether natural or management induced, what factors might have influenced what you are seeing on the ground? Fire or flood. Heavy downpour or hailstorm, a complete rainfall failure. Have stock been in the paddock recently or not for months? Did you create herd effect with an attractant?
4. *Change or no change.* If adverse or no changes have occurred, what is the underlying cause. Positive changes are important but more important are the adverse changes. Carefully consider the tools you have used. For example: Even though you increased stocking rate and reduced the size of paddocks, the soil is still bare and capped and no change occurred. Maybe grazing and animal impact were trumped by partial rest.
5. *Proposed actions.* What are you going to change next year? In most cases, the action will require a tool other than the one that led to the adverse change or a modification in how you applied the tool. Continuing with the above example: To determine how to overcome the inadvertently applied tool of rest, you would look to animal impact but not as you applied it last time. Plan to increase it significantly. The applicable management guidelines are stock density and herd effect. To increase herd effect: 1) amalgamate herds or 2) use an attractant or 3) strip-graze. To increase stock density: 1) lease or purchase additional animals or 2) amalgamate herds or add temporary fence to further reduce paddock size. In the end, assume you are wrong and continue to monitor.

File your summary forms with their data forms and photos in such a way that they won't get separated.

Biological Monitoring Analysis - Basic
(Use one per transect)

Property _____ Transect/Plot Numbers_____ Photo No's _____

Date_____ Examiner(s)_____

1. What are we trying to achieve in the area surrounding this transect?

```
Community Dynamics:

    Water Cycle:

    Mineral cycle:

    Energy flow:
```

2. What progress have we made his year, compared to last?c

```
Community Dynamics:

    Water Cycle:

    Mineral cycle:

    Energy flow:
```

3. What natural or management factors might have influenced what we are seeing on the ground?

4. If adverse changes have occurred or no change, where change was planned: What is the underlying cause—what tools have we applied, and how have we applied them?

5. What are we going to change in this next year to keep our land moving toward the future landscape described in our holistic goal?

Comprehensive Monitoring

This type of monitoring is especially essential if you are managing public lands or working for an absentee owner.

Locating Transects: Pick areas that are either typical or where you want to produce a lot of change. The more uniform the land, the fewer transects you will need. Three to five are usually sufficient on uniform ranches but varying terrain requires more.

You will be gathering data from random points within each transect area. There are three requirements for point sampling: 1) adequate number of points; 2) randomness in choosing points; 3) points that really are points.

Choose the points by throwing a dart backward over your shoulder. Sample up to 100 points at each site (makes it easy to express the data as percentages).

The Monitoring Forms: The Comprehensive Biol. Monitoring *Data* form has three divisions: First, record what the dart point hit (covered or bare ground); Second, note what is within a 6 inch circle; Third, record the distance from the sample point to the nearest perennial plant and the information about that plant.

The Comprehensive Biol. Monitoring *Summary* form is for recording totals and averages of the data from the monitoring data sheets.

The Comprehensive Biol. Monitoring *Analysis* form is where you record your analysis of the results.

Equipment Needed

- Camera (digital)
- Monitoring data sheets
- Clipboard and pencils
- Measuring tape
- Fishing weight on a length of cord (if in brushy country)
- Steel posts (3 per transect)
- Optional short marker post (1 per transect and only required the first time when marking boundaries)
- Bright colored darts with 2 to 3 inch tips.
- Monitoring instructions

Identifying and Marking the Transect Area: Locate permanent starting points for each transect which will consist of up to 100 random points. The starting point remains the same from year to year.

Boundaries are defined with a camera. Standing at the permanent starting point, face a fixed feature and center it in the camera's viewfinder. The left and right extremes of the camera's field of view define the triangular boundaries. Mark the starting point and the boundaries with steel posts. Record the location on a map and write down directions for getting there. GPS coordinates can be used but should be backed up by information on paper.

Recording Transect Data

1. ***Set up the Comprehensive Biological Monitoring form (3 pages).*** Fill in the top (except for photo numbers) of each page.
2. ***Take Photos.*** Write the transect number and date on a piece of paper (or board) and place it within the camera's field of view. Stand against the starting point with the feature centered in the viewfinder, 1/3 sky and 2/3 foreground and take the picture. Use the same camera each year or at least be consistent with the lens setting.

Now get a more detailed view of the soil surface. There are two ways to do this: a) remain where you are and point the camera down at a 45 degree angle or b) For a closer, more detailed view, walk (five to ten yards) away from the market toward the fixed feature (use the same distance for all sites). This gets you away from the influence of livestock concentrating around the marker. Drive in a short marker with a distinctive top. Take the photo from waist high with the camera pointed straight down with the head of the marker centered in the viewfinder.

When you print the photos, number them and record their numbers at the top of the monitoring data sheets.

3. ***Throw the Dart.*** From the permanent starting point, throw the dart backward over your shoulder anywhere within the transect area. (Short tosses are easier to find.)

When you finish one sample point, move on to the next. Take care to spread points so they cover the entire monitoring site. If the terrain is very uniform, 50 throws should be ample.

It is useful to walk around the area before throwing the dart and record names of familiar plants because, with random dart throws, some species may not show up in the data.

4. ***Record the Data.*** Enter the data to the monitoring data sheet a row per throw.

At the Dart Entry Point: If the dart does not stick, sight straight down over the point and check the appropriate column (Follow along with the data sheet on the following pge:

- Bare soil and rock—self-explanatory
- Litter 1—new un-decayed litter (leaves, sticks, dung).
- Litter 2—deeper decaying litter
- Plant Base—the area covered by the root crowns of plants

Basal area in most bunchgrass communities rarely exceeds 5 to 15%.

Biological Monitoring Data - Comprehensive										
Property: _____	Transect # _____	Photo Nos. _____	Date: _____ Examiners: _____							

		At Dart Entry Point (Must Check One)	6 inch Circle Around the Point		Describe the Nearest Perennial					
		What Dart Point Hit	Yes	Soil Surface (Must Check One)	Evidence of (Check if "Yes")	What it is (Must check one)	Its Habitat (Check one)	Its Age (Check one)	Its form (Must check one)	Species if Known	Other Comments
Throw Number	Bare Soil	Litter 1	Litter 2	Rock	Plant Base	Canopy Above Point	Mature	Immature	Recent	Broken	Covered	Animal Sign I,W,B,S,L	Annuals Present	Soil Movement	Grass C,W or Y	Rush or Sedge	Forb	Shrub	Tree	Distance (in) to it	Dry	Middle	Wet	Seedling	Young	Mature	Decadent (Dying)	Resprout	Normal	Overrested	Overgrazed	Overbrowsed	Dead		
1																																			
2																																			
3																																			
4																																			
5																																			
6																																			
7																																			
8																																			
9																																			
10																																			
11																																			
12																																			
13																																			
14																																			
15																																			
16																																			
17																																			
18																																			
19																																			
20																																			
21																																			
22																																			
23																																			
24																																			
25																																			
26																																			
27																																			
28																																			
29																																			
30																																			
31																																			
32																																			
33																																			
Total																																			

Individual areas are hard to distinguish with sod-forming grasses so you have to develop a standard and stick with it. One way is to record a basal hit only if the point falls into a *living* clump. If not, record liter or bare ground and measure the distance to the nearest live shoot of sod-forming grass.

If the dart lodges in above ground vegetation, use the plumb bob to find the point on the surface directly below.

If the point is immediately under a canopy of any sort (leaf or branch) that will slow a raindrop, check "canopy above point."

In a 6 inch circle centered on the entry point:

1. ***Record soil surface conditions:*** Check the column that best describes these conditions.

    - Mature capping—Algae, lichen or moss covers surface
    - Immature capping—surface has been broken in the past but is still hard enough to inhibit change
    - Recent capping—recently sealed over by rain
    - Broken capping—bare surface has been recently broken but has not yet formed a distinct cap
    - Covered—most of the raindrops falling within the circle would land on living or dead plant material rather than mature capping or bare ground

Note: If the surface within a large number of circles is sod bound with low successional species, record it as "covered" but make a note in the "other comments" column.

2. ***Record evidence of large or small animal activity.*** Tracks, droppings, burrows, mounds, worm castings or actual sightings. Note in "animal sign" column: I = insect; W = worm; B = bird; L or S = large or small. If you can identify the species, note it in "other comments."

3. ***Note if annuals are present.*** Check "annuals present" if an annual grass or forb is within the circle. If you know the name, record it in "other comments."

4. ***Record any evidence of erosion.*** Check "soil movement" if there are any signs within the circle—plants or rocks on pedestals, water flow patterns, litter banks, siltation in low points or splash patterns.

The Nearest Perennial Plant

Note: When specifically trying to create perennial grassland, you might want to locate the nearest perennial *grass* plant.

- *Note the type of plant* in the appropriate column:

    - *Grass*—C or W for Cool or Warm Season or Y for green year-round.
    - *Rush or Sedge.* Sedges resemble grasses but have solid, angular (vs. rounded) stems; rushes have hollow, pithy stems.
    - *Forb.* Any flowering plant that does not develop woody stems (e.g. a legume like clover)
    - *Shrub or tree.* Fix your own definition according to species in your area.
- *Measure distance from the dart point* to the base of the perennial plant and record under "Distance to It." *Note:* In pure stands of dense grass, record the point strike with a check and a letter for the species and then measure the distance to the nearest perennial of *another* species.
- *Record habitat.* Indicate whether "dry," "middle," or "wet."
- *Record age.* Check "seedling," "young," "mature," "decadent (dying)," or "re-sprout" (from an existing plant).
- *Record the form.* This is the plant shape under the influence of grazing or rest. "Normal"—the plant is vigorous with evidence of seed production, tillering and branching and a lack of old, stale growth; "over-rested"; "overgrazed"; "over-browsed"; and "dead." If the form fits none of these, note that in "other comments."
- *Note the species.* Identify it if you can. If not, simply record "unknown."

When done, subtotal the checks (or letters) in each column. If you took 100 points, the combined sub totals (except "distance to it") are automatically percentages.

Summarizing the Data

Fill in the top of the monitoring summary form. If the area was uniform and you only recorded 50 points, calculate percentages first by multiplying by 2.

Transfer the following from the data sheets:

- *Plant name or species.* In the "plant name" column, list the species you could identify within each plant type and indicate how many of each to the right under "No-%." Lump any that you could not identify under "unknown."
- *Soil surface.* In "cover & capping" record the total of the relevant columns. In "Evidence of - %" the number of Is, Ws, Bs, Ss and Ls in the "animal sign" column of the data sheets. Enter the sum of the subtotals from the monitoring sheets for "annuals" and "soil movement."
- *Nearest perennial.* Enter the numbers for "cool season," "warm season," and "year-round" under the "characteristics - %" column then the totals for rushes, forbs, shrubs and trees. Average the "distance to it" column of the data sheets and record it in the "average distance" row of the summary sheet.

Refer to the Summary form on the following Page

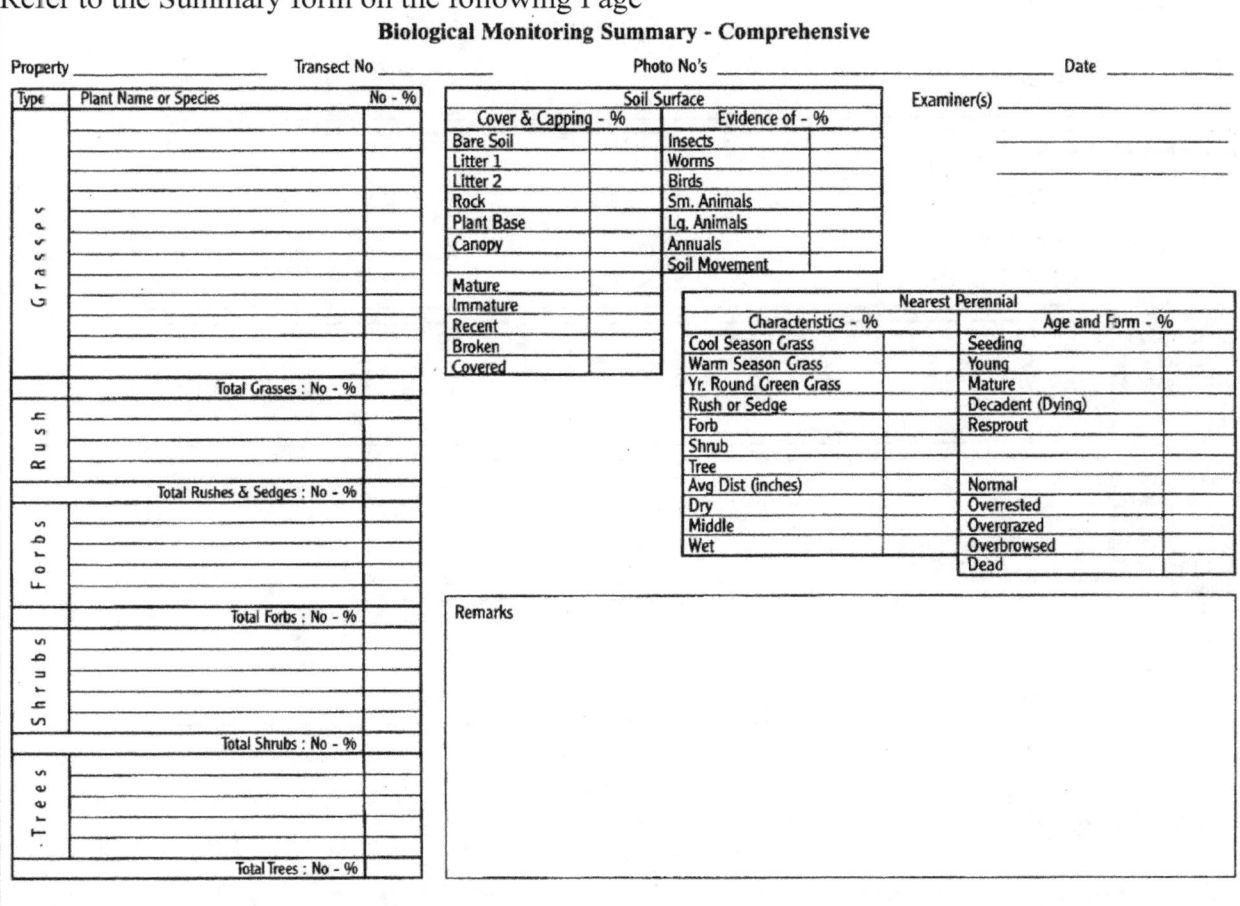

Analyzing the Data:

Fill in the identification information at the top of the monitoring analysis form. First year data is baseline. In subsequent years your analysis will be the basis for a great many decisions. The questions asked are the same as those in the basic monitoring procedure. (Follow along with the Analysis Form on the following page.)

Future Landscape Description

1. *What are we trying to achieve?* Answer in terms of the 4 ecosystem processes and indicate if you are trying to produce a specific landscape (e.g. brush). The answer will likely remain the same for every year.

2. *Progress Check:* What progress have we made this year over the last? Rate the status of the 4 ecosystem processes. Note specific changes in community dynamics, water and mineral cycles and energy flow:

Biological Monitoring Analysis - Comprehensive

Property _____ Transect No _____ Photo No's _____ Date _____

1. What are we trying to achieve in the area surrounding this transect?
   - Community Dynamics:
   - Water Cycle:
   - Mineral cycle:
   - Energy flow:

2. What progress have we made this year, compared to last?
   - Community Dynamics:
   - Water Cycle:
   - Mineral cycle:
   - Energy flow:

3. What natural or management factors might have influenced what we are seeing on the ground?

4. If adverse changes have occurred or no change, where change was planned. What is the underlying cause—what tools have we applied, and how have we applied them?

5. What are we going to change in this next year to keep our land moving toward the future landscape described in our holistic goal?

2.a. Community Dynamics

- *Cover*: An increase in bare ground and rock indicates a decline; Litter (especially Litter 2) is a precondition for advancing succession. Basal cover indicates plant density.
- *Capping*: Mature capping is a sign of decline. Covered soil (except sod-bound) signals an advance. Immature, recent and broken capping indicate an advance if they increase while mature capping decreases.

- *Animal Sign:* Becomes more meaningful over the years as it changes. An "infestation" reflects a lack of balance but is not necessarily bad. Certain species dominate certain levels of succession but cause little damage when predators are diverse and abundant.
- *Annuals Present:* If you are attempting to convert an annual to a perennial grassland, these should decrease relative to perennials.
- *Soil Movement:* Erosion and bare ground go hand in hand with low successional communities.
- *Plant type and species*: Consider the general diversity, the proportion of annuals and the presence of high-successional plants. Changes in cool- and warm-season grasses are significant in terms of diversity.
- *Average distance:* An index of plant density. Closely spaced plants hold soil and litter and provide soil cover. Comparisons to later monitoring provide sensitive indicators of trend. Expanding bare areas and falling plant density are signs of decline.
- *Plant age and form:* The youngest class determines the community of the future.

2.b. Water Cycle

- **Cover:** Liter or basal cover enhances the water cycle. Excessive grazing of dry vegetation or consumption of litter off of the ground, results in adequate cover despite minimal overgrazing during the growing season.
- *Capping:* Primary indicator of a non-effective water cycle
- *Animal Sign:* borrowing animals affect the water cycle. More activity is usually better but the evaluation is subjective.
- *Soil Movement:* A direct indicator of a poor water cycle.
- *Plant type and species:* plants with varying root depths get their water at different levels so diversity reflects health. Species associated with good or bad soil aeration may indicate changes in the water cycle. Perennials improve the water cycle.
- *Average distance:* Changes in plant density may precede changes in litter retention and water cycle.
- *Plant age and form:* Water cycle changes explain why seedlings do or don't survive. Overgrazed and over-rested plants provide less litter for soil cover.

2.c. Mineral Cycle

- *Cover:* Litter (especially litter 2) is a direct measure of improvement.
- *Capping:* inhibits the mineral cycle.
- *Animal Sign:* All creatures that live in the soil contribute to mineral cycling.
- *Soil Movement:* Any loss of soil is a loss of minerals from the system.
- *Plant type and species:* Look for plants of varying root depth and diversity of species. Deep-rooted perennial grasses should be increasing.
- *Average distance:* Greater plant density implies more cycling.
- *Plant age and form:* An indicator of growth activity. Dying and slowly decaying plants slow the cycle.

2.d. Energy Flow

- *Cover:* Bare ground or rock indicates the worse possible.
- *Capping:* Always indicates a loss of potential energy flow.

- *Animal Sign:* Better energy flow supports more activity.
- *Soil Movement:* Erosion means bare ground. Thus, it affects energy flow.
- *Plant type and species:* Broad-leafed, middle-type plants indicate more energy flow. A mix of cool- and warm-season plants means solar energy will be converted over a longer period and thus better energy flow. An increase in perennial grasses relative to annuals, increases energy flow.
- *Average distance:* Tighter density means a greater volume of leaf which means more sunlight energy is harvested.
- *Plant age and form:* Green leaf area determines energy flow. Dead or dying plants don't convert much energy.

3. **Influencing Factors:** *What natural or management factors might have influenced what we are seeing on the ground?* Fire or Flood; Weather (heavy downpour or hailstorm); stock in the paddock recently; you recently created herd effect with an attractant; etc.

4. **Change or No Change:** *What is the underlying cause—what tools have you applied and how?* If you have a poor water cycle because of capping and absence of litter, insufficient animal impact (partial rest) is probably the cause. Think about each tool, how it was applied, and how it might have affected the four processes. (Remember: Rest produces opposite effects in brittle vs. non-brittle environments.) The following will help you determine answers and explain their underlying cause.

Interpreting Your Results—Five Scenarios

The scenarios in the table reflect changes that may have occurred since the previous monitoring. Think through each of them and the reasoning behind what happened.

Scenario	Bare Ground	Mature Capping	Litter	Broken	Plant Spacing	Over-Grazing	Over-Rest	Weeds or Woody Plants
1	Decrease	Decrease	Increase	Increase	No change	None	None	Increase
2	Increase	Decrease	Decrease	Increase	Decrease	None	None	No change
3	Decrease	Decrease	Increase	Increase	Decrease	None	None	Decrease
4	No change	No change	Increase	No change	No change	None	Increase	Increase
5	Decrease	Decrease	Increase	Increase	Decrease	Yes	Yes	Increase

Scenario 1: Decreasing bare ground, increasing broken soil and decreasing mature capping all suggest animal impact. Litter increase could be animal impact. But, unchanged plant spacing while oxidizing grass has disappeared (no signs of over-rest), implies that animal impact is still too low (partial rest too high). The land is improving but slower than planned.

Scenario 2: No overgrazing or over-resting is good. Mature capping decreasing and broken soil increasing indicate some animal impact. But, increasing bare ground and decreasing litter are not good. The probability is that the problem is overstocking (indicated by disappearance of litter because they *have eaten it*).

Scenario 3: Decreasing bare ground, increasing broken soil, decreasing mature capping, increasing litter, closer plant spacing and increasing numbers of grass plants are all in line with planned changes.

Scenario 4: No changes in bare soil, mature capping, broken soil, or plant spacing all suggest inadequate animal impact or too much partial rest. Combine with an increase in over-rested plants and new weeds; these indicate that the stocking rate is too low.

Scenario 5: Decreasing bare ground, mature capping and plant spacing and increasing litter all suggest effective animal impact. However, some plants showing signs of overgrazing while others are over-resting and weeds are increasing indicate that the manager either has been rotating the livestock or has not been paying attention to recovery periods and daily growth rates.

What each tool produces in terms of the 4 processes:

Fire

- *Community dynamics:* Exposes the soil and inhibits establishment of new plants that require litter, moisture and low temperature fluctuations. Stimulates woody species by initiating sprouting (only a few woody species are killed by fire). Short term increase in diversity but, when repeated, it reduces diversity. Can produce mosaic patterns (edge effect) and thus biological diversity.
- *Water cycle:* Fire reduces effectiveness because it exposes soil and destroys litter.
- *Mineral cycle:* Speeds it up in the short-run but, if frequently repeated, slows the cycle in the long-term.
- *Energy flow:* Short term—fire produces an increase in energy flow in oxidizing, over-rested grasslands. But, soil exposure leads to less effective mineral and water cycles and energy flow is reduced in the long-term.

Rest—non-brittle environments. The most powerful tool for restoring biodiversity in *non-brittle* environments.

- *Community dynamics:* Communities develop levels of great diversity and stability
- *Water and mineral cycles:* Builds and maintains high levels of effectiveness.
- *Energy flow:* Reaches a high level.

Rest—very brittle environments. When applied continuously, biodiversity and soil cover are damaged.

- *Community Dynamics:* Gradual chemical oxidation and physical weathering. Communities decline and greater simplicity and instability ensue.
- *Water and mineral cycles:* Less effective
- *Energy flow* declines significantly.

Animal impact (high).

- *Community dynamics:* Promotes advancement of biological communities on bare, gullied and eroding ground. In dense grassland, maintains the community at the grassland level and prevents a shift to a woody community.
- *Water and mineral cycle:* Improves water and mineral cycles.
- *Energy flow:* Because it builds community complexity and improves water and mineral cycles, it also improves energy flow as a consequence.

Animal impact (low).

- *Community dynamics* Low animal impact or partial rest produces bare ground. It allows plant spacing to increase and may allow grasslands to proceed toward woody communities with bare or algae-lichen covered ground.
- *Water and mineral cycles:* Reduces these.

- *Energy flow.* Reduces this below its potential.

Grazing.

- *Community dynamics:* maintains grass root vigor, soil life and structure and retards shifts toward woody species.
- *Water and mineral cycles:* Enhances both by maintaining healthier root mass, increasing microorganism and aeration and plants with more shoots and lives—thus providing litter.
- *Energy flow:* Increases both above and below ground by preventing old, oxidizing blockages and promoting vigorous root and leaf growth. Healthier root systems support microorganisms and other life underground.

Overgrazing.

- *Community dynamics:* Reduces litter and soil cover, damages grass roots and fosters shifts away from grassland to woody plant communities. Soil-enhancing legumes disappear.
- *Water and mineral cycles:* Reduces both by exposing soil and limiting liter production. Reduces grass roots which fosters soil compaction.
- *Energy flow.* Cuts energy flow by reducing plant roots and exposing the soil surface.

Living organisms. When used as positive "tools" each specific case must be analyzed. There is danger that organisms used will subsequently find other hosts. In some cases, they have been used to eradicate populations that could have been better dealt with through grazing and high animal impact.

Technology. The thousands of available technologies cannot be easily broken down into categories on which to base general tendencies. Use common sense. Pesticides reduce biodiversity and thus adversely affect all four ecosystem processes.

5. *Proposed Actions: What are we going to change next year to keep moving toward our goal?* Take action right away. In most cases you will need a tool other than the one that led to the adverse change—or a modification in how you applied it. Determine which guidelines apply and then decide how to use the tool.

Again using the example of a poor water cycle brought on by insufficient animal impact. The two management guidelines that apply most directly are stock density and herd effect.

To improve density, you can increase fencing, amalgamate herds or increase stocking rate.

To improve herd effect, you can train animals to respond to attractants or strip-graze at ultra-high density.

File your summary and analysis forms together with their data forms and photos.

A Note on Shortcuts: If for some reason you can't complete the whole process in a given year, the following are important:

- Fixed-point photos
- Soil surface information
- Thorough notes on observations (questions 1-5 on the monitoring analysis forms)